Developments in Geotechnical Engineering

Series editors

Braja M. Das, California State University, Henderson, CA, USA
Nagaratnam Sivakugan, James Cook University, Townsville, QLD, Australia

This series on Geotechnical Engineering has been initiated to provide an integrated platform for publishing in all areas of geotechnics and their applications to engineering problems. The series focuses on the topics that have garnered recent interest from a large research audience, including, but not limited to, saturated and unsaturated soils, expansive soils, geosynthetics, marine geotechnics, and soil improvement. This series includes monographs, textbooks, and references in the general area of Geotechnical Engineering. The books in this series will serve researchers, practitioners, and students alike.

More information about this series at http://www.springer.com/series/13410

K. Ilamparuthi · R. G. Robinson
Editors

Geotechnical Design and Practice

Selected Topics

Springer

Editors
K. Ilamparuthi
Department of Civil Engineering,
College of Engineering Guindy
Anna University
Chennai, Tamil Nadu
India

R. G. Robinson
Department of Civil Engineering
Indian Institute of Technology Madras
Chennai, Tamil Nadu
India

ISSN 2364-5156 ISSN 2364-5164 (electronic)
Developments in Geotechnical Engineering
ISBN 978-981-13-0504-7 ISBN 978-981-13-0505-4 (eBook)
https://doi.org/10.1007/978-981-13-0505-4

Library of Congress Control Number: 2018940875

Printed on acid-free paper

This Springer imprint is published by the registered company Springer Nature Singapore Pte Ltd.
part of Springer Nature
The registered company address is: 152 Beach Road, #21-01/04 Gateway East, Singapore 189721, Singapore

Foreword

The Indian Geotechnical Society (IGS) was started as the Indian National Society of Soil Mechanics and Foundation Engineering in the year 1948, soon after the Second International Conference on Soil Mechanics and Foundation Engineering held at Rotterdam. The Society was affiliated to the International Society in the same year, and since then, it has strived to fulfil and promote the objectives of the International Society. In December 1970, the name 'Indian Geotechnical Society (IGS)' was adopted.

The Society conducted several symposia and workshops in different parts of India since its inception in 1948. It was in the year of 1983, the Indian Geotechnical Society organized its first annual conference IGC 1983 in Indian Institute of Technology Madras.

Several local chapters of the Society were established over the years, and gradually, the annual conferences were held in different cities under the leadership of the respective local chapters. The conferences were well utilized as the venue for showcasing the research works and the case studies on geotechnical engineering and geo-environmental engineering, and the papers presented during the deliberations were being published as conference proceedings volume.

The responsibility of organizing the annual conference of 2016 was taken up by IGS Chennai Chapter ,and the conference was held during 15–17 December 2016. Eminent professors and practitioners were invited to present keynote and theme lectures during this conference. These keynote and theme lectures comprise this volume 'Geotechnical Design and Practice—Selected Topics' under the series 'Developments in Geotechnical Engineering'.

The Chennai Chapter of Indian Geotechnical Society placed on record its acknowledgement of the efforts put forth by Springer in bringing out this book for the benefit of the geotechnical engineering community.

Chennai, India

Prof. A. Boominathan
Chairman, Indian Geotechnical Conference 2016
Department of Civil Engineering
Indian Institute of Technology Madras

Preface

The Indian Geotechnical Society has been organizing national-level annual conference since 1962. The Indian Geotechnical Conference 2016 (IGC 2016) was conducted at IIT Madras in collaboration with the Chennai Chapter of Indian Geotechnical Society, Indian Institute of Technology Madras and College of Engineering, Guindy, Anna University, Chennai, from 15 to 19 December 2016. The theme of the conference was 'Geotechnology Towards Global Standards'. To fulfil the objectives of the conference, it was proposed to cover various disciplines of geotechnical engineering, and therefore, experts from various parts of the globe were identified and invited to share their knowledge by presenting keynote lectures and invited papers.

There were 12 keynotes and 17 invited speakers, and their presentation covered a wide spectrum of geotechnical engineering from fundamental properties of soils to remote sensing applications including large-scale test to understand passive response of skewed bridge abutments. The speakers are well-known researchers in the fields of soft ground improvement, seismic response of retaining structure using SSI principles, unsaturated soils, etc. The deliberations of eminent researchers generated useful discussions among the practising geotechnical engineers, consultants, researchers and academicians. It was decided to bring together an edited book based on the talks delivered by the keynote and invited speakers.

With this view, a book on 'Geotechnical Design and Practice—Selected Topics' was compiled which contains 22 chapters covering properties of soils, unsaturated soil mechanics, ground improvement, liquefaction and seismic studies, soil–structure interaction and stability analysis of man-made and natural slopes.

We believe that the contents of the book on the recent developments in the said areas of geotechnical engineering will be a useful source of reference for practising engineers and researchers apart from providing insights for future research. The editors are grateful to all the speakers of keynote and invited papers for sparing their valuable time in preparing the source material and sharing their knowledge with participants during the sessions of the conference. Finally, we the editors of this volume would like to express our sincere thanks to the organizers of IGC 2016 for the wonderful opportunity of editing the volume.

Chennai, India

Dr. K. Ilamparuthi
Dr. R. G. Robinson
General Editors

Contents

About the Editors

Dr. K. Ilamparuthi is a professor in the Department of Civil Engineering, College of Engineering Guindy, Anna University, Chennai, with more than three decades of teaching and research experience. He obtained his Ph.D. from the Indian Institute of Technology Madras, India. His areas of research include ground improvement, soil–structure interaction, anchor foundations, slope stability analysis and finite element analysis. Over the course of his career, he has published over 200 research papers in journals and conference proceedings and has guided 9 Ph.D. students and 145 M.E. students. He is actively involved in the administrative running of the College of Engineering, Guindy, and has held various posts including dean in-charge, chairman of the Faculty of Civil Engineering and head of department.

He is an active consultant involved in several major projects. He has received numerous prestigious awards for his research and his consultation work, as well as for the Commonwealth fellowship. He worked at the University of Liverpool, UK, for a period of 13 months from October 1999 to November 2000 and has visited several countries such as UK, Belgium, Australia, Hong Kong, USA, Middle East and Singapore on various assignments.

Dr. R. G. Robinson is a professor in the Geotechnical Engineering Division in the Department of Civil Engineering at IIT Madras, India. He obtained his Ph.D. from the Indian Institute of Science Bangalore, India, and has pursued postdoctoral research at the National University of Singapore, in addition to visiting University of Wollongong in Australia for a period of 6 months under the Endeavour India Fellowship. His research interests include physical modelling, soil characterization and ground improvement with special reference to soft clays and expansive soils. He has published over 100 research papers in journals and conference proceedings and has guided 8 Ph.D. and 15 Master's students. He and his students received the IGS-Mr H. C. Verma Diamond Jubilee Award for Innovative Instrument Design from IIT Madras.

Quality Assurance Studies for Ground Improvement Projects

Anand J. Puppala, Tejo V. Bheemasetti and Bhaskar C. S. Chittoori

1 Introduction

Soils are inherently heterogeneous in nature with wide distribution of their physical and mechanical properties (Phoon and Kulhawy 1999). When poor subsoil conditions are encountered, engineers often improve the soil properties by choosing suitable ground improvement techniques. Comprehensive state-of-the-art ground improvement techniques for different soil types were discussed in different papers and keynote lectures. (Mitchell 1981; Puppala and Perez 2009). Terashi and Miki (1999) have provided a detailed overview for the selection of ground improvement technique for a specific project based on the subsoil conditions. Broadly, the most commonly used ground improvement techniques can be classified into two categories: mechanical stabilization and chemical stabilization.

Mechanical stabilization such as deep dynamic compaction, vibro-compaction, stone columns, pre-loading, deep dynamic compaction are few prominent techniques which stabilize the soils by rearranging the soil particles (Mitchell and Zoltan 1984). Chemical stabilization technique refers to enhancing the soil properties by treating with additives such as lime, cement, fly ash, and other by-products (Bredenberg et al. 1999; Rathmayer 2000). The chemical stabilization technique became widely accepted ground improvement technique for treating highly expansive clayey soils which is the main focus of this paper.

A. J. Puppala (✉) · T. V. Bheemasetti
Civil Engineering Department, The University of Texas at Arlington, Arlington, USA
e-mail: anand@uta.edu

T. V. Bheemasetti
e-mail: tejovikash.bheemasetti@mavs.uta.edu

B. C. S. Chittoori
Civil Engineering Department, Boise State University, Boise, USA
e-mail: bhaskarchittoori@boisestate.edu

K. Ilamparuthi and R. G. Robinson (eds.), *Geotechnical Design and Practice*,
Developments in Geotechnical Engineering,
https://doi.org/10.1007/978-981-13-0505-4_1

Since the twentieth century, new changes in the ground improvement techniques have taken place. This is mainly due to the innovations in geotechnical equipment and construction procedures. Along with new ground improvement techniques, considerable advancements have taken place in quality control and quality assurance methods. The bi-directional arrangement of osterberg cells (o-cell) in traditional load tests has improvised the quality assurance tests in deep foundations.

The pre- and post-standard penetration tests and cone penetration tests more accurately represents the variation in the strength properties before and after the implementation of ground improvement techniques (Sondermann and Weher 2004; Raju 2010). Researchers have also proposed and implemented new quality assurance and quality control methods for different ground improvement techniques (Puppala et al. 2004; Madhyannapu et al. 2010). In this paper, quality assurance studies performed on a pipeline bedding material and at a pavement infrastructure were discussed. In these studies, new design and construction procedures were implemented. The following subsequent sections present the details of these studies.

2 QA/QC in Pipeline Infrastructure

A 150-mile water pipeline was proposed to provide additional water supplies to Dallas/Fort Worth Metroplex area. The proposed pipeline is 2.74 m (9 ft.) in diameter and passes through the six different geological formations namely, Eagle Ford, Kemp, Wills, Neylandville, Ozan, and Wolfe.

Extensive borehole investigations were performed along the six (6) geological formations to determine the subsurface profile and soil properties. The subsurface profile mainly consists of clays with low to high plasticity characteristics. Of all, the Eagle Ford geological region depicted high plasticity characteristics with plasticity index of 37 and liquid limit of 62. The average swell pressure is observed to be 40.7 psi (280.6 kPa) and average volumetric strain is observed to be −19.7%. The native high plasticity soils possessed a serious threat to the pipeline project as the soils exhibited expansive swell/shrink behavior. In order to provide a proper bedding material to the pipeline material, the controlled low strength material (CLSM) was considered as a viable option.

CLSM is a self-compacted, cementitious material, which was widely known as flowable fill until American Concrete Institute Committee 229 documented its name as CLSM (ACI 1994). The CLSM is primarily used as a backfill material in lieu of compacted backfill and has become a popular material for projects such as void fill, foundation support, bridge approaches, and conduit bedding (Folliard et al. 2008). CLSM, with different additives such as cement and fly ash, has been demonstrated, by many researchers, to be an effective bedding material for pipelines due to the material's self-compacting behavior and strength performance (Rajah et al. 2012; Boschert and Butler 2013). In this study, the CLSM was prepared using native plasticity soils and Type I/II Portland cement. The utilization of native soils will provide sustainable engineering solution by minimizing the project cost and by reducing the negative impact on the environment (Puppala and Hanchanloet 1999; Chittoori et al. 2012; Puppala et al. 2012).

2.1 *Pipeline Installation and CLSM Mix Design*

In order to study the feasibility of CLSM prepared with native clayey soil as a pipeline bedding material, it was first tested on a 152.4 m (500 ft.) long test section. This test section was also referred to as prove-out section. Ten (10) pipeline sections of 15.2 m (50 ft.) were used to construct the full prove-out pipeline length of 152.4 m (500 ft.).

After excavating the soils and laying the pipeline, the native plasticity soils from test site were used in the place of aggregates in preparing the CLSM mix. The mix design of CLSM constitutes of cement, water and native plasticity soil. Based on laboratory studies conducted by Raavi (2012), 4% Type I/II Portland cement was selected to prepare the CLSM mix for the present test section. Table 1 provides the basic soil properties on native soils collected from the test site. Soil classification tests were performed based on Unified Soil Classification System (USCS). Sieve and hydrometer tests were conducted as per ASTM D 422 and Atterberg's Limit Tests (Liquid Limit and Plastic Limit) were conducted as per ASTM D 4318 standards. Specific gravity test was conducted in accordance with the ASTM D 854.

The CLSM prepared at the site was poured into the trench in two different layers. First, the CLSM was poured to 30% height of the pipeline. Due to its flowability nature the CLSM was spread around the pipeline and later it was finally poured up to 70% height of the pipeline. The stiffness monitoring studies were conducted at various elapsed curing periods to evaluate its material behavior. The quality assurance studies conducted were discussed in the subsequent sections.

2.2 *Quality Assurance Studies*

In this research study, an attempt was made to check the feasibility of CLSM prepared using native plasticity soils as a pipeline bedding material. Laboratory

Table 1 Soil properties and CLSM mix design

Basic properties	
Percent passing No. 200 sieve	73
USCS classification	CL
Liquid limit, LL	24.4%
Plastic limit, PL	14.3%
Plasticity index, PI	10.1%
Specific gravity, G_s	2.62
Cement used in CLSM	4%

studies were performed to check the strength enhancement of CLSM samples prepared using native plasticity soils. The test results and analysis demonstrated the enhancement in strength that satisfies the criteria for pipeline bedding material (Puppala et al. 2007; Raavi 2012, Chittoori et al. 2014). However, the assurance of strength and stiffness characteristics in the field is often challenging. The in situ stiffness of CLSM will be influenced by several factors such as construction methodology, environmental factors, and existing soil conditions. Past researchers have attributed several water pipeline failures due to lack of proper inspection (Kienow and Kienow 2009). In this research study, a quality assurance method, which is a combination of Spectral Analysis of Surface Waves (SASW) technique and Geostatistics, was proposed to ascertain the stiffness of pipeline bedding material.

2.2.1 Spectral Analysis of Surface Waves (SASW)

SASW technique was developed in early 1980s based on the dispersive characteristics of surface waves (Nazarian and Stokoe 1984; Stokoe et al. 1989). Based on the nature of investigation, different types of tools can be employed for generating the seismic waves and recording the wave forms. In this research study, SASW tests were performed at seventeen (17) test sections along the prove-out section, comprising of at least one (1) test section for every 9.1 m (30 ft.). Five (5) test points (South 2, South 1, Center, North 1, and North 2) were selected at each section to determine the variation in stiffness of CLSM bedding material across the pipe as shown in Fig. 1.

SASW tests were performed with a geophone spacing of 0.61 m (2 ft.). This distance was selected so that the stiffness profile of CLSM beneath the pipeline (0.45 m ~ 1.5 ft. thickness) can be obtained. The recorded waveforms were analyzed to determine the stiffness of the CLSM across the pipeline. Since, the thickness of pipeline is only 0.025 m (1 in.); the results obtained are direct indication of CLSM across the pipeline. However, a higher thickness of the pipeline must be carefully analyzed as the waves can directly pass through the pipeline.

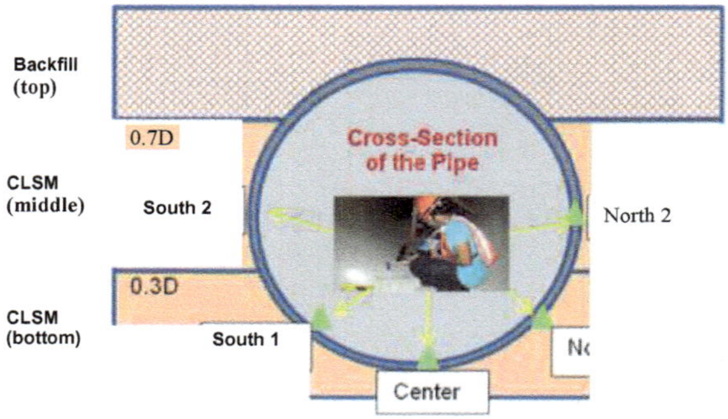

Fig. 1 Cross-section of test section inside pipeline

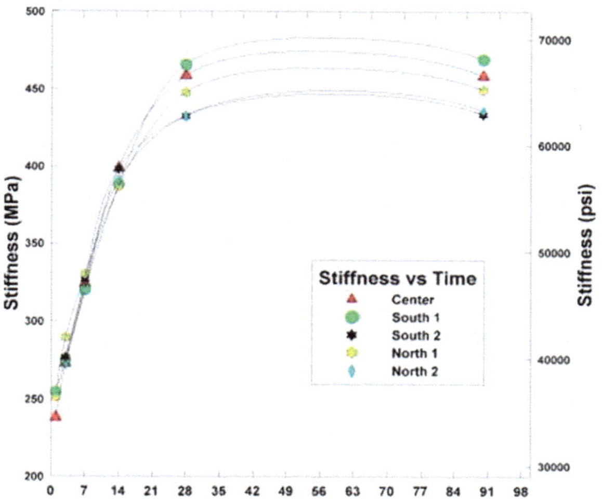

Fig. 2 Stiffness enhancement at section 1066-25

Figures 2 represent the stiffness profile of CLSM with respect to different curing periods in days. The five (5) test points at each section have depicted an increase in the stiffness values with an increase in curing period. A close observation of the results indicates that, there has been a consistent increase in stiffness values from day 1 to 28, but the rate of increase in stiffness of CLSM from day 28 to 90 was not significant. This suggests that CLSM had attained most of its strength by the end of 28 days curing period.

2.3 Geostatistical Studies

The geostatistical studies were employed in this research to analyze the spatial variability of stiffness measurements obtained from SASW tests. Geostatistics is a separate branch of statistics which deals with spatial analysis of the data sets (Isaaks and Srivastava 1989). The application of geostatistics was used mainly in the mining industry to predict the location of ore by describing the probability distribution of the existing ore locations (Krige 1951). In this study, geostatistical analysis was performed on SASW data obtained from the inside of the pipe as well as that obtained from the surface. The two main important steps that were performed in geostatistical analysis is modeling the spatial variability in stiffness values through variogram and performing predictions using kriging analysis.

2.3.1 Variogram Analysis

The variogram or semi-variogram γ (h) is a traditional geostatistical analysis tool used to model the spatial variability in the data sets. Mathematically, it is defined as one-half of the average squared differences between the x and y coordinates of each pair of points in the h-scatter plot (Isaaks and Srivastava 1989).

In this study, variogram analysis was performed to identify and model the spatial variability present in the stiffness measurements. This was performed by constructing the variogram plots, where variogram values were calculated and plotted against the corresponding lag distance values. Figure 3 represents the experimental variogram plots for the stiffness measurements obtained after 28 days curing period. The variogram values are plotted on the y-axis which is dependent on the lag distance values that are plotted on the x-axis.

The range: 24; sill: 600 and nugget: 200 are three important characteristics that are modeled in the variogram plot. The range represents the spatial continuity of the data sets and sill represents the level at which the variogram reached asymptotic value, whereas the nugget represents the variability in the small distances. In this study, the nugget behavior is due to the variability in the five tests points across the section. The blue solid exponential line in the variogram plot represents the spatial variability model developed for stiffness measurements. Similar analysis is conducted for all the stiffness measurements obtained at different curing periods. The spatial variability in the stiffness measurements obtained at different curing periods is modeled using the nugget effect and exponential model. The developed spatial variability models were incorporated in the kriging analysis for predicting the stiffness measurements of CLSM at untested locations.

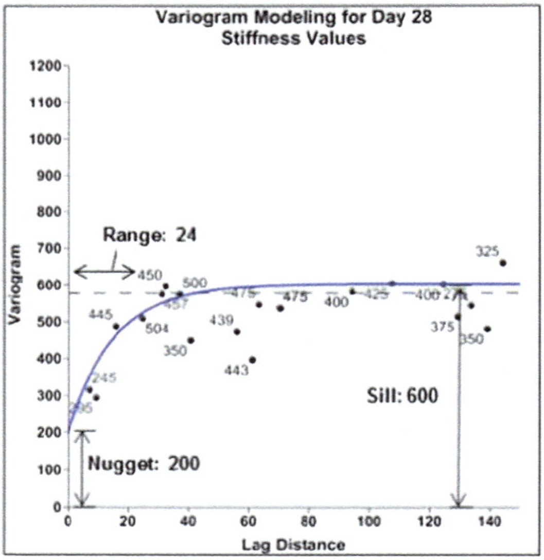

Fig. 3 Variogram plot for stiffness measurements

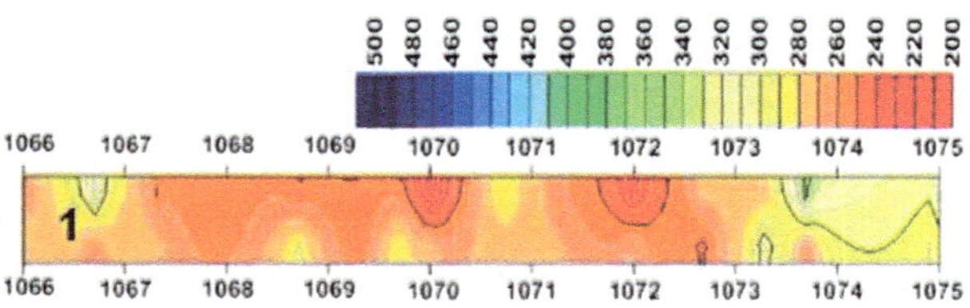

Fig. 4 Variation of stiffness measurements using kriging analysis

2.3.2 Kriging Analysis

The kriging analysis was used in this study to predict the stiffness measurement along the entire pipeline using the known measurements and spatial variability models. Kriging provides the best linear unbiased estimates to predict the values at unknown locations (Armstrong 1994). This is because of its ability to reduce the error variance of the predicted values. The spatial variability models developed earlier for the stiffness values were used along with the kriging algorithms to obtain the weights of the neighboring values around the untested location. Figure 4 represents the day 1 and day 28 stiffness variation of the CLSM along the pipeline in the plan view. The horizontal axis represents the 10 sections 1066 to 1075 of 152.4 m (500 ft.) length.

The color index presented at the top of Fig. 4 represents the stiffness values of CLSM in MPa. It can be observed that after 1 day of curing period, the CLSM depicted a stiffness of 200–280 MPa and on day 28 the stiffness has significantly increased to more than 360 and 400 MPa. Also, it can be observed that the stiffness development after each curing period is uniform along the prove-out section. This ascertains the quality of the CLSM prepared in the field using the native soil and self-compaction effort the CLSM.

2.4 Surface Testing for Quality of Stiffness Assessments

The main objective of this task is to determine the variation of stiffness of the subsurface layers including CLSM and backfill across and beneath the pipeline. This analysis was performed by conducting both SASW tests from the surface and geostatistical analysis. Several trial and error procedures were adopted for the surface testing as the wave interaction with the pipeline creates an erroneous wave forms which could not be analyzed. After repeated trails and with interpolations a final design test layout as shown in Fig. 5 was implemented. It should be noted that 0.91 m (3 ft.) spacing in the geophone provides a seismic wave to a depth of 1.83 m (6 ft.). In view of the above perspective the geophone spacing are kept at 0.91 m (3 ft.), 1.22 m (4 ft.), 1.52 m (5 ft.) and 2.44 m (8 ft.) to avoid impact of pipeline.

SASW tests were conducted from the surface using geophones as receivers. Tests were conducted at three locations within 30.5 m (100 ft.) stretch with different distant

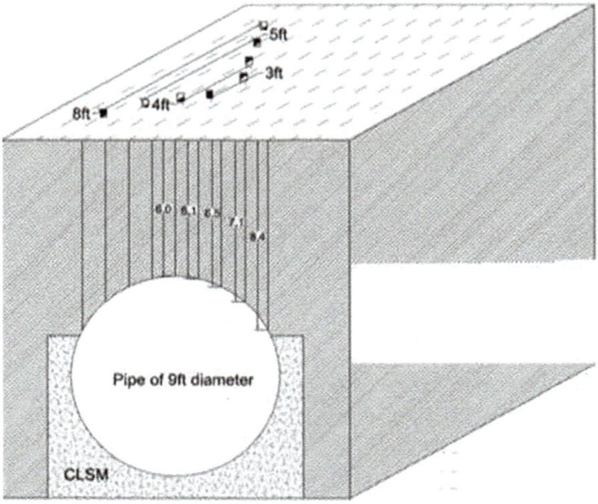

Fig. 5 SASW test grid layout from surface

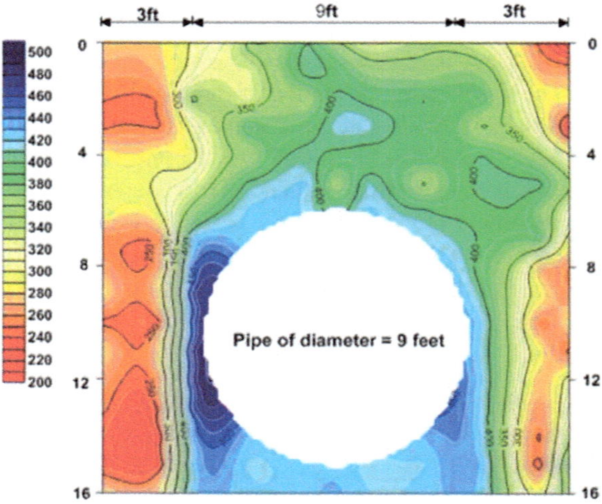

Fig. 6 Subsurface stiffness variation

spacing. Necessary precautions were taken to ensure the wave forms generated are not from any field construction activity by adjusting the sensitivity of the equipment. Geostatistical analysis was conducted on the stiffness measurements obtained from the surface tests. The spatial variability model developed for the stiffness measurements determined after 28 days curing period was incorporated with the kriging analysis to predict the variation of stiffness measurements along the subsurface layers.

Figure 6 presents the variation of stiffness layers from the surface. The vertical axis in Fig. 6 represents the depth of the subsurface profile. The hollow circle in the figure represents the 2.74 m (9 ft.) diameter pipeline.

From Fig. 6, it can be observed that the material around the pipeline indicates high stiffness value, corresponding to CLSM. The stiffness values decreases with an increase in the distance from the pipeline depicting the transition from CLSM to soil. The soil above the CLSM depicts high stiffness values which can be due to continuous compaction effort performed in the field. Also, a clear pattern of high stiffness values are observed on the left side of the pipeline. This can be attributed to the side where the CLSM was poured into the trench.

2.5 Summary and Conclusions

In the current study, the CLSM prepared using native plasticity soil was used as a bedding material to the pipeline of length 152.4 m (500 ft.). An attempt was made to ascertain the stiffness enhancement of CLSM using both seismic non-destructive test method and Geostatistics. Seventeen test sections were chosen along the prove-out section to monitor the stiffness behavior. The test results and analysis depicted that the stiffness of CLSM has increased significantly with increase in curing period. After 28 days curing period, the stiffness has reached to a constant value. The SASW test was successfully utilized to determine the stiffness of CLSM and geostatistics could predict the stiffness variation of the CLSM along the prove-out pipeline section.

3 QA/QC Studies in Deep Soil Mixing Technique (DSM)

Deep soil mixing (DSM) is a ground improvement technique mainly used to enhance strength and stiffness properties of soft clays, loose sands, peaty soils, and problematic soils such as expansive clays. Often the stabilization is performed using the lime or cement, of which the former is more specifically used to reduce the hydraulic conductivity of treated soils due to flocculation (Rathmayer 1996; EuroSoilStab 2002). Over the years, researchers and practitioners have demonstrated the successful use of DSM technique over wide variety of projects (Rathmayer 1996; Porbaha 1998; Bruce 2001). In this research study, the DSM technique is implemented to stabilize the expansive soils and mitigate the shrink and swell behavior. One of the key parameter that influences the overall performance in this technique is quality assurance studies.

Several factors such as soil type, binder type and concentration, binder–water ratio, curing conditions, mixing methods and construction practice play a key role in influencing the performance of a DSM project (Madhyannapu et al. 2010). Especially, it is highly necessary to ensure that the design strength parameters achieved in the laboratory and field conditions are same. This paper presents the quality assurance studies performed on expansive soils treated using DSM

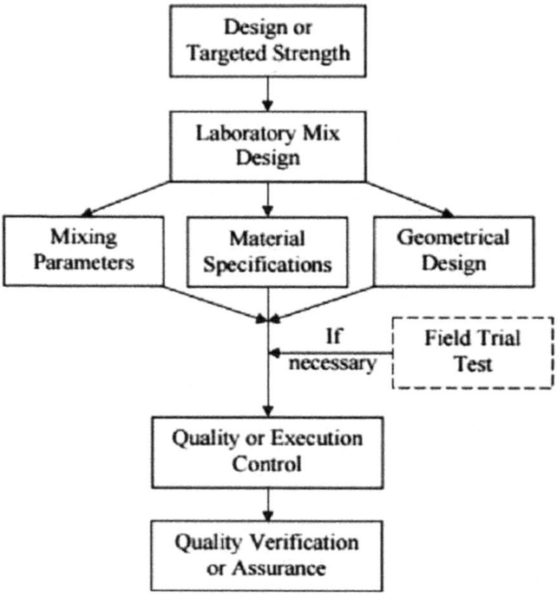

Fig. 7 Typical QA/QC procedure (modified after Coastal Development Institute of Technology 2002 and Usui 2005)

technique. Figure 7 presents the modified version of typical QA/QC procedure adopted in DSM technique.

In order to evaluate the DSM technique for expansive subgrade soils, two field test sites located in Fort Worth, Texas were selected. QA/QC studies were performed to evaluate the effectiveness of DSM technique in mitigating the swell-shrink behavior of expansive soils. Both test sites comprises of expansive clay subsoils with medium to high plasticity characteristics. Soils from two test sites were treated using 25% lime and 75% cement with a binder dosage of 200 kg/m^3 and w/b ratio of 1.0. During the field construction of DSM columns, the wet samples were grabbed at different depths to assess the strength and stiffness parameters of the soils. Similar field composition of DSM samples was also prepared in the laboratory with a unit weight close to the field samples. Both laboratory and field samples were subjected to a curing period of 14 days.

Stiffness measurements were performed using the bender element tests and strength tests were performed using unconfined compressive strength tests. Also, to evaluate the effectiveness of this treatment technique for expansive behavior, free swell tests and linear shrinkage bar tests were performed on both laboratory and field specimens. The test results depicted that the shear wave velocity for moderate and high PI treated soils at both curing periods of 7 and 14 days varied from 24.6 (170 MPa) to 43.6 psi (301 MPa) and 25.5 (176 MPa) to 46.7 psi (322 MPa), respectively. The improvement in stiffness of treated soils, when compared to the control soils, was approximately 4–7 times for Site 1 and 5–9 times for Site 2. Table 2 presents the field to laboratory strength and stiffness ratios.

Table 2 Strength and stiffness ratios

Site	$G_{max,field}/G_{max, lab}$	$q_{u,field}/q_{u,lab}$
1	0.43–067	0.67–0.70
2	0.56–0.65	0.83–0.86

From Table 2, it can be inferred that the field strength values were 20–30% lower than the laboratory measurements, whereas the field stiffness measurements were 40% lower than the laboratory measurements. Also, in order to evaluate the in situ stiffness measurements, field tests were performed using natural gamma logging, down hole P-wave velocity and the SASW testing. Natural gamma-ray measurements and Downhole P-wave velocity tests were performed in the cased boreholes at each site. SASW tests were performed along two parallel lines (to balance the effect of wave paths relative to the DSM columns for shallow depths) in the treated area and one line in the untreated area (outside the treated area) at each site.

The natural gamma logging tests were performed to detect the variations in the natural radioactivity originating from changes in concentrations of the trace elements Uranium (U) and thorium (Th) as well as changes in concentration of the major rock forming element potassium (K). The downhole tests were performed to estimate the seismic wave velocity profile. In the DSM column at two test sites, P-wave velocity ranges from 1080 to 1140 m/s, whereas in the untreated areas the P-wave velocity averages around 780 m/s which is significantly lower than the global average value in the treated area. The SASW tests were performed to evaluate the shear wave velocity profile in the treated and untreated areas.

In order to evaluate the performance of the DSM technique in expansive clays at in situ condition, periodic measurements were taken in the field for two years. The Gro-Point moisture probes were installed in field to measure the moisture content. The average moisture content in site 1 varied from 13 to 30% over a depth of 14 ft. (4.3 m), whereas in site 2 the moisture content varied from 24 to 30% over a depth of 14 ft. (4.3 m). In order to measure the swell and shrink movements, the horizontal and vertical inclinometer casings were installed in both treated and untreated sections of sites 1 and 2. It was observed that the range of surface movements of Site 1 is 0.07–0.74 in. (0.17–1.87 cm) and 0.12–0.63 in. (0.3 to 1.6 cm), respectively, and of Site 2 for Phases I and II are 0.06–0.12 in. (0.15–0.3 cm) and 0.01–0.25 in. (0.025–0.63 cm), respectively. The site 2 depicted less movement due to increase in the treatment area ratio when compared to site 1. Also, periodic measurements were obtained using both downhole testing and SASW testing. It was observed that in both the tests, the initial measurements taken in the first year is considerably higher than the measurements taken over the next two year period. However, the treated sections depicted considerably higher values when compared to the untreated sections.

Overall, the comparisons between the field and laboratory test results indicated that the stiffness ratio $G_{max,field}/G_{max,lab}$ for Site 1 and Site 2 specimens varied between 0.43–0.67 and 0.56–0.65, respectively. The strength ratios ($q_{ucs,field}/q_{ucs,lab}$)

for Site 1 and Site 2 varied from 0.67 to 0.70 and 0.83 to 0.86, respectively. Both stiffness and strength ratios indicate that the field stiffness and strength values are 40 and 20–30% lower, respectively, when compared to the laboratory treatments. The P-wave velocities of the treated soil column zones exhibited higher values than those recorded in the untreated soils. Also, the same measurements for the three consecutive yet different years showed a decrease in the P-wave velocities.

4 Conclusions

Quality assurance studies play a key role in evaluating the new techniques that are implemented in geotechnical engineering projects. In this research paper, two quality assurance techniques performed in two different case studies were discussed. In both the studies, the quality assurance studies are highly challenging demanding for new techniques. In the first case study, the implementation of SASW testing along with geostatistical theories provided a real time assessment of the stiffness of CLSM with respect to curing period. In the second study, the quality assurance procedures along with advanced statistical analysis have provided the actual performance of the DSM technique.

Acknowledgements The authors would like to acknowledge Mr. David Marshall, Ms. Shelly Hattan, and Mr. Richard S. Williammee for their help and excellent cooperation throughout the research.

References

American Concrete Institute, Committee 229, Controlled Low-Strength Materials (CLSM). 1994. *ACI 229R-94 Report*.

Armstrong, M. 1994. Is research in mining geostats as dead as a dodo? In *Geostatistics for the next century*, ed. R. Dimitrakopoulos, 303–312. Dordrecht: Kluwer Academic.

Boschert, J., and J. Butler. 2013. CLSM as a pipe bedding: Computing predicted load using the modified Marston equation. *ASCE Pipelines* 2013: 1201–1212.

Bredenberg, H., G. Holm, and B.B. Broms (eds.). 1999. Dry mix methods for deep soil stabilization. In *Proceedings of the international conference on dry methods for deep soil stabilization*, Stockholm, 358 p.

Bruce, D. 2001. *An introduction to the deep mixing methods as used in geotechnical applications. Volume III. The verification and properties of treated ground.* Report No. FHWA-RD-99-167, US Department of Transportation, Federal Highway Administration.

Chittoori, B., A.J. Puppala, R. Reddy, and D. Marshall. 2012. Sustainable reutilization of excavated trench material. In *Geocongress 2012*, ASCE, Reston, VA, 4280–4289.

Chittoori, B., A. Puppala, and A. Raavi. 2014. Strength and stiffness characterization of controlled low-strength material using native high-plasticity clay. *Journal of Materials in Civil Engineering* 26 (6): 0401400.

EuroSoilStab. 2002. *Design guide soft soil stabilization*. Project No. BE 96-3177, Ministry of Transport Public Works and Management.

Folliard, K.J., David, Trejo, S.A. Sabol, and D. Leshchinsky. 2008. *Development of a recommended practice for use of controlled low-strength material in highway construction*, NCHRP 597 Report, 5–39.

Isaaks, E.H., and R.M. Srivastava. 1989. *An introduction to applied geostatistics*. New York: Oxford University Press.

Kienow, K., and K. Kienow. 2009. Most pipeline failures can be prevented by proper inspection. *Pipelines* 2009: 182–197. https://doi.org/10.1061/41069(360)18.

Krige, Danie G. 1951. A statistical approach to some basic mine valuation problems on the Witwatersrand. *Journal of the Southern African Institute of Mining and Metallurgy* 52 (6): 119–139.

Madhyannapu, R.S., A.J. Puppala, S. Nazarian, and D. Yuan. 2010. Quality assessment and quality control of deep soil mixing construction for stabilizing expansive subsoils. *ASCE, Journal of Geotechnical and Geoenvironmental Engineering* 136 (January, 1), 119–128.

Mitchell, J.K. 1981. State of the art—Soil improvement. In *Proceedings of the 10th ICSMFE*, Stockholm, vol. 4, 509–565.

Mitchell, J.K., and V.S. Zoltan. 1984. Time dependent strength gain in freshly deposited or densified soil. *Journal of Geotechnical Engineering, ASCE* 110 (11): 1559–1576.

Nazarian, S., and K.H. Stokoe II. 1984. In situ shear wave velocities from spectral analysis of surface waves. In *Proceedings of the 8th world conference on earthquake engineering*, vol. III, 31–38. Englewood Cliffs, New Jersey: Prentice-Hall, Inc.

Phoon, K., and F.H. Kulhawy. 1999. Characterization of geotechnical variability. *Canadian Geotechnical Journal* 36: 625–639.

Porbaha, A. 1998. State-of-the-art in deep mixing technology, Part I: Basic concepts and overview of technology. *Ground Improvement* 2 (2), 81–92.

Puppala, A, J., and S. Hanchanloet. 1999. Evaluation of chemical treatment method (sulphuric acid and lignin mixture) on strength and resilient properties of cohesive soils. In *Proceedings of the 78th Transportation Research Board annual meeting*. Washington, DC: Transportation Research Board.

Puppala, A.J., A. Porbaha, V. Bhadriraju, and E. Wattanasanthicharoen. 2004. In situ test protocols for quality assessments of deep mixing columns. In *Geo-Trans 2004*, ASCE Geotechnical Special Publication No. 126, Los Angeles, 2004, 1429–1438.

Puppala, A.J., P. Balasubramanyam, and R.S. Madhyannapu. 2007. Experimental investigations on properties of controlled low-strength material. *Ground Improvement Journal* 11 (3): 171–178.

Puppala, A.J., and D. Perez. 2009. New ground improvement techniques for expansive soils. In *12th geotechnical conference*, Bogota, Colombia, Keynote Lecture, October 22–25, 2009.

Puppala, A.J., S. Saride, and R. Williammee. 2012. Sustainable reuse of limestone quarry fines and RAP in pavement base/subbase layers. *Journal of Materials in Civil Engineering*, 418–429. https://doi.org/10.1061/(asce)mt.1943-5533.0000404.

Raavi, A. 2012. *Design of controlled low strength material for bedding and backfilling using high plasticity clay*. M.S. thesis, University of Texas, Arlington, TX, 118.

Rajah S., M. McCabe, and J. Plattsmier. 2012. Classification and specification of bedding and backfill for buried pipelines. ASCE Pipelines 2012: Innovations in Design, Construction and Maintenance, 940–951.

Raju, V.R. (2010) Ground improvement—Applications and quality control. In *Proceedings of Indian geotechnical conference*, 121–131.

Rathmayer, H. 1996. Deep mixing methods for soft soil improvement in the Nordic countries. In *Proceedings of the 2nd international conference on ground improvement geosystems, grouting and deep mixing*, 14–17 May, Tokyo, 2, 869–877.

Rathmayer, H. (ed.). 2000. Grouting, soil improvement including reinforcement. In *Proceeding for the 4th international conference on ground improvement geosystems*, Helsinki, 570 p.

Sondermann, W., and W. Weher. 2004. Round Improvement, 2nd ed, 57–92. Spon Press, Taylor and Francis Group.

Stokoe II, K.H., G.J. Rix, and S. Nazarian. 1989. In situ seismic testing with surface waves. In *Proceedings of the 12th international conference on soil mechanics and foundation engineering*, Rio De Janiero, 331–334.
Terashi, M., and H. Miki. 1999. Importance of prediction in ground improvement. In Prediction and performance of ground improvement, 1–10. Practitioners Series No. 11. Japanese Geotechnical Society.

Liquefaction Screening—Non-plastic Silty Sands and Sands

S. Thevanayagam, U. Sivaratnarajah, V. Veluchamy and Q. Huang

1 Introduction

1.1 Soil Liquefaction and Screening

Current liquefaction screening techniques rely on knowledge from extensive laboratory research conducted on liquefaction resistance of clean sands, and extrapolations of observed field performances during past earthquakes (NCEER 1997; Youd et al. 2001). Such observations have been documented in the form of normalized penetration resistances (SPT $(N_1)_{60}$, CPT q_{c1N}) (Seed et al. 1983; Robertson and Wride 1998), and shear wave velocity (v_{s1}) (Andrus and Stokoe 2000) versus cyclic stress ratio (CSR) induced by the earthquakes, corrected for a standard earthquake magnitude of 7.5, for many soil deposits where occurrence or non-occurrence of liquefaction were recorded during the earthquakes (Fig. 1). For liquefaction screening applications, the cyclic resistance ratio (CRR) of a soil deposit, applicable for number of cycles and frequency content relevant for a standard earthquake magnitude of 7.5, with a known value of $(N_1)_{60}$, q_{c1N} or v_{s1} for the site is obtained from a demarcation line drawn between the past field-observation-based data points which correspond to liquefied deposits and those that did not liquefy in Fig. 1. This is denoted as $CRR_{7.5}$. This $CRR_{7.5}$ value is

S. Thevanayagam (✉) · U. Sivaratnarajah · V. Veluchamy · Q. Huang
Department of Civil Engineering, University at Buffalo, Buffalo, USA
e-mail: theva@buffalo.edu

U. Sivaratnarajah
e-mail: umaipala@buffalo.edu

V. Veluchamy
e-mail: vijayakr@buffalo.edu

Q. Huang
e-mail: qhuang4@buffalo.edu

K. Ilamparuthi and R. G. Robinson (eds.), *Geotechnical Design and Practice*, Developments in Geotechnical Engineering,
https://doi.org/10.1007/978-981-13-0505-4_2

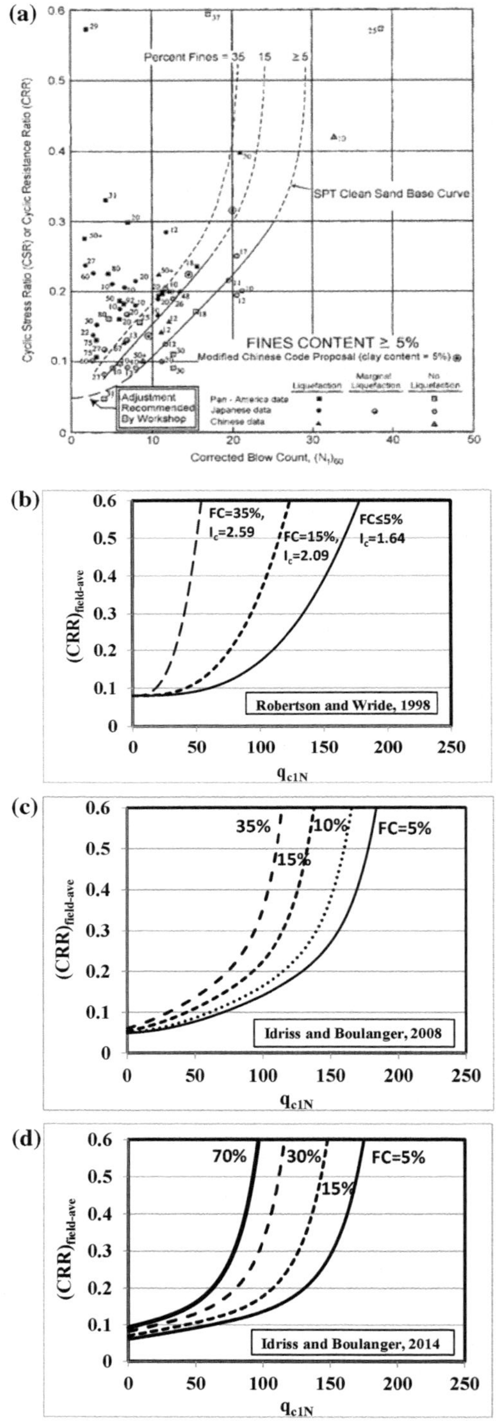

Fig. 1 Silt content-dependent liquefaction screening charts

compared against the anticipated CSR for that deposit due to a design earthquake of the same magnitude to determine whether or not that deposit would liquefy. Factor of safety (FS) against liquefaction is defined as FS = $CRR_{7.5}$/CSR. If a different design earthquake magnitude is expected, the $CRR_{7.5}$ is multiplied by a magnitude scaling factor (MSF) to obtain the CRR applicable for the deposit for the relevant number of cycles, frequency content, etc., of the design earthquake. The factor of safety is FS = MSF * $CRR_{7.5}$/CSR.

Practitioners and researchers have observed that the $CRR_{7.5}$ determined in this manner depends on silt content of the soil for a given $(N_1)_{60}$, q_{c1N} or v_{s1}. For the same values of $(N_1)_{60}$, q_{c1N} or v_{s1}, silty sands show higher $CRR_{7.5}$ than for sands (Fig. 1). Demarcation lines have been developed for silt content less than 5, 15, and 35%, respectively. In the case of cone penetration, an equivalent cone index I_c which is considered to represent the apparent effects of silt content has been introduced to obtain these demarcation lines (Fig. 1b). These different demarcation lines are used to determine most probable $CRR_{7.5}$ for a deposit, depending on the silt content (or I_c) of the soil deposit. While this approach has been routinely used in practice, a rational understanding of this procedure has been lagging.

As evident in Fig. 1b–d, these silt content-dependent screening demarcation lines in the charts continue to evolve based upon the engineers' experience (e.g., Robertson and Wride 1998; Idriss and Boulanger 2008; Boulanger and Idriss 2014). One of the main sources of uncertainty is that the effects of presence of silt among sand grains on liquefaction resistance (CRR), q_{c1N}, and the interrelationship between silt content, silt soil characteristics (such as permeability k, compressibility m_v, and coefficient of consolidation c_v), CRR, and q_{c1N} remain largely unexplored and unknown. Successful application of the current screening charts to other sites calls for significant judgment on the part of the practitioner. A rational approach to liquefaction screening requires a fundamental understanding of the above '*silt-phenomenon*' observed in these charts (viz. the interrelationship between silt content, k, m_v, CRR, q_{c1N}), and development of revised/new charts that incorporates this new understanding for it to be applied successfully with a high degree of confidence for all sites.

1.2 Recent Research and Key Results

Recent research shows that the silt-phenomenon affecting the liquefaction screening relationship (CRR vs. q_{c1N}) is as follows (Thevanayagam and Martin 2002; Thevanayagam et al. 2006, 2016a, b).

- Silt particles in a silty sand do not fully contribute to the intergranular contacts and associated force chains. Therefore, they do not fully contribute to the cyclic strength ratio CRR and cone resistance q_{c1N} as much as they contribute to the density of the silty sand. For this reason, a silty sand may appear dense and can have low void ratio. But it may not be as high resistant to liquefaction and as high resistant to cone penetration as a clean sand at the same void ratio.

- Silt particles in a silty sand matrix contribute to the reduction in porosity and pore opening size and therefore contribute to a reduction in k and c_v. Therefore, a cone penetration process in a saturated silty sand produces sizable excess pore pressure around the cone that does not dissipate rapidly (due to low k and c_v) during penetration, and as a result contribute to a reduction in q_{c1N}. On contrary, a clean sand is highly permeable, and therefore cone penetration produces low excess pore pressure, due to rapid dissipation of pore pressures around the cone (due to large k and c_v) of the sand, leading to somewhat higher q_{c1N}. Furthermore, the pore pressure regime and its dissipation rates during penetration in sands and clays have been shown to depend on cone diameter (d) and penetration velocity (v) (Thevanayagam et al. 2006; Schneider et al. 2007). Therefore, they also affect the measured q_{c1N}.
- The above combined effects influence the CRR-q_{c1N} relationship to a different degree for silty sand than for a clean sand. This phenomenon is not just a function of the amount of silt content only, but rather highly depends on k and c_v as well, if v and d are kept the same.
- A refined and more rational liquefaction screening method could be developed taking into account the effects of c_v, which can be measured using piezocone tests, on q_{c1N} (Thevanayagam et al. 2016a, b) and appropriately accounting for the effect of silt content on CRR, based on intergranular void ratio concept (Thevanayagam et al. 2002). Progress to date (presented below) indicates that it is highly likely there exists a rational relationship between CRR-q_{c1N}-T (where $T = vd/c_v$).

This paper presents a summary of (a) intergranular void ratio concept, (b) recent advances on understanding of the influences of non-plastic silt content and intergranular contact density on CRR, (c) effects of silt content on k and c_v, (d) recent findings on the effect of intergranular contact density, k, c_v, v, and d on q_{c1N} of sand and silty soils, and (e) possible relationships between CRR, q_{c1N}, and T. Preliminary simplified design charts for liquefaction screening of non-plastic silty sands and sands is presented.

2 Silt Content—Intergranular Contacts and Mechanical Properties

2.1 *Intergranular Void Ratio Concept*

The equivalent intergranular void ratio concept (Thevanayagam 1998; Thevanayagam and Mohan 2000; Thevanayagam et al. 2002) proposes that mechanical properties such as liquefaction resistance (CRR), steady state strength, shear wave velocity, stress–strain characteristics, and the likes of soils are influenced by intergrain contact density of a soil, among other factors. Silty sand and sand at the same global void ratio e are not expected to have the same intergrain contact density. Sand–silt mixes and host sand are expected to show similar

Fig. 2 Intergranular void ratio concept

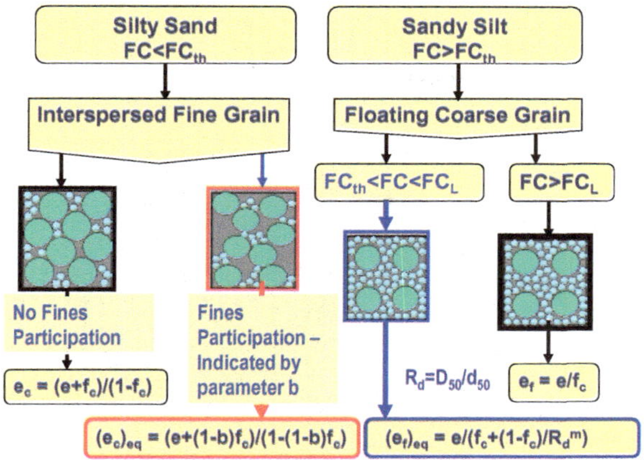

mechanical behavior if compared at a same contact density index. To this effect, a soil matrix classification system (Fig. 2) was developed (Thevanayagam et al. 2002). Two approximate measures for contact density, namely equivalent intergranular void ratio indices, $(e_c)_{eq}$ and $(e_f)_{eq}$, were introduced for soils at silt content (FC) less than a threshold silt content FC_{th} and more than FC_{th}, respectively. $(e_c)_{eq}$ and $(e_f)_{eq}$ have been defined as

$$(e_c)_{eq} = [e + (1 - b)\mathrm{fc}]/[(1 - (1 - b)\mathrm{fc})] \tag{1a}$$

$$(e_f)_{eq} = \left[e/\left(\mathrm{fc} + (1 - \mathrm{fc})/R_d^m\right)\right], \tag{1b}$$

where fc = fines content by weight, R_d = ratio of the d_{50}'s of the host sand and silt in the soil mix; b and m are soil parameters depending on gradation and grain size characteristics of the soil such as uniformity coefficient of coarse grain soil (C_{us}) and fine-grained soil (C_{uf}) in the soil mix (Kanagalingam and Thevanayagam 2006a, b; Thevanayagam 2007a, b). In order to broaden the application of intergranular concept, an equivalent intergranular relative density $(D_{rc})_{eq}$ has been defined as

$$(D_{rc})_{eq} = \left((e_{max})_{HS} - (e_c)_{eq}\right)/\left((e_{max})_{HS} - (e_{min})_{HS}\right), \tag{2}$$

where HS = host sand.

A preliminary analysis of the dependence of b-parameter in Eq. (1a) on grain size and gradation characteristics indicate an approximate relationship for b in terms of R_d, C_{us} and C_{uf} as shown in Fig. 3, based on database for more than 30 soils obtained from the literature (Veluchamy 2012). It may be more appropriate to consider $R_d = D_{85}/d_{15}$. The literature does contain other correlations as well for b with soil gradation, including silt content (e.g., Ni et al. 2004; Kanagalingam and Thevanayagam 2006a, b; Rahman et al. 2008).

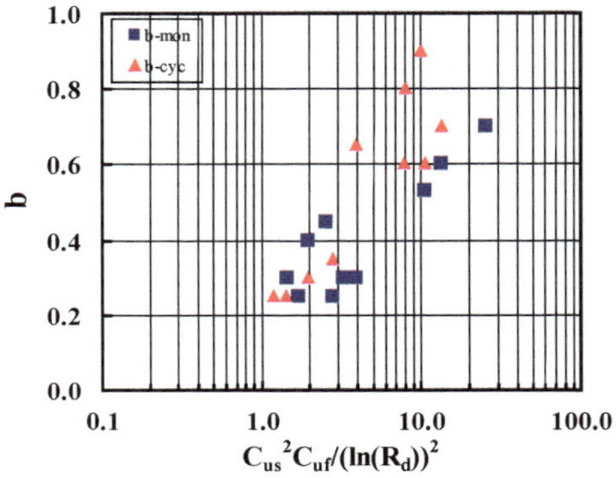

Fig. 3 Parameter b versus soil grain and gradation characteristics

A review of the literature indicates that the intergranular void ratio concept appears to broadly capture the essence of the effects of silt content on mechanical properties of soils, including cyclic strength, for practical purposes (Thevanayagam 1998; Thevanayagam et al. 2002; Thevanayagam 2007b; Veluchamy 2012). This is illustrated below, for cyclic strength of soils, using a few example data sets.

2.2 *Cyclic Resistance (CRR) Versus* $(\mathrm{D}_{rc})_{eq}$

Effect of non-plastic silt content on cyclic resistance has been the subject of research and much controversy in the early 80s until recently. A large data base (Veluchamy 2012) has been recently compiled based on available data in the literature on the effects of non-plastic silt content on undrained cyclic resistance of silty soils. Figure 4 shows the number of cycles (N_L) required to reach liquefaction

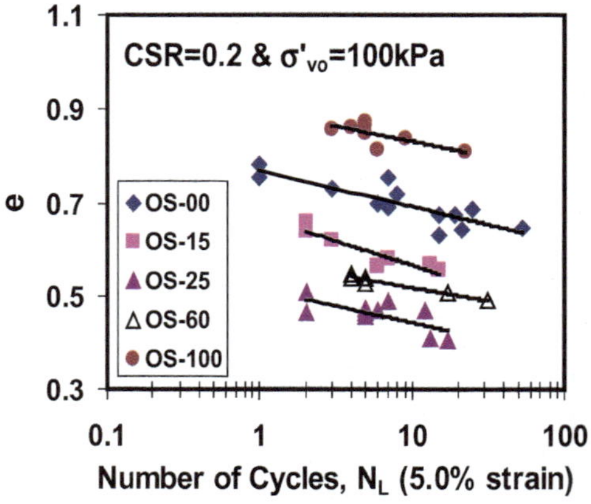

e versus N_L

Fig. 4 Effect of silt content on liquefaction resistance versus e

versus e for Ottawa sand/silt mix obtained from undrained cyclic triaxial tests conducted at a cyclic stress ratio (CSR) of 0.2 and initial confining stress of 100 kPa. The specimens were prepared by mixing Ottawa sand (D_{50} = 0.25 mm) with a non-plastic silt (d_{50} = 0.01 mm) at different silt contents (0–100% by weight). OS-15 in this figure refers to sand–silt mix at 15% silt content. At the same e, liquefaction resistance of silty sand decreases with an increase in silt content. Beyond a transition silt content of about 20–30%, the trend reverses and liquefaction resistance increases with further increase in silt content. Similar observations have been widely reported in the literature (e.g., Zlatovic and Ishihara 1997; Koester 1994; Polito and Martin 2001; Carraro et al. 2003, Cubrinovski and Rees 2008). However, for the reasons described earlier in Sect. 2, it is not appropriate to compare the liquefaction resistance of silty soil with that of clean sand using e as depicted in Fig. 4.

To examine the utility of the intergranular void ratio concept (Fig. 2), the data set shown in Fig. 4 was reorganized into two groups, one for silt content (FC) less than a threshold value (FC_{th}) and the other for silt content exceeding FC_{th}, and replotted against $(e_c)_{eq}$ and $(e_f)_{eq}$, instead of e in the x-axis. Figure 5a, b show these

Fig. 5 Effect of silt content on liquefaction resistance versus $(e)_{eq}$

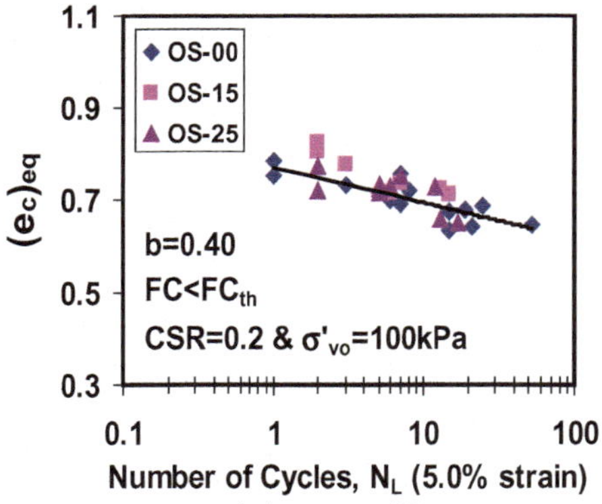

(a) $(e_c)_{eq}$ versus N_L

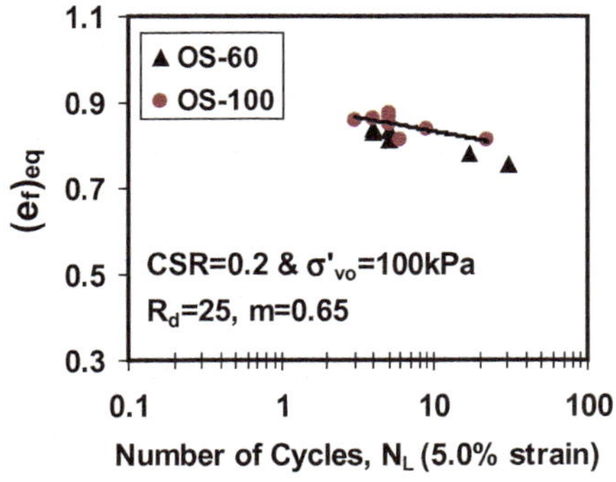

(b) $(e_f)_{eq}$ versus N_L

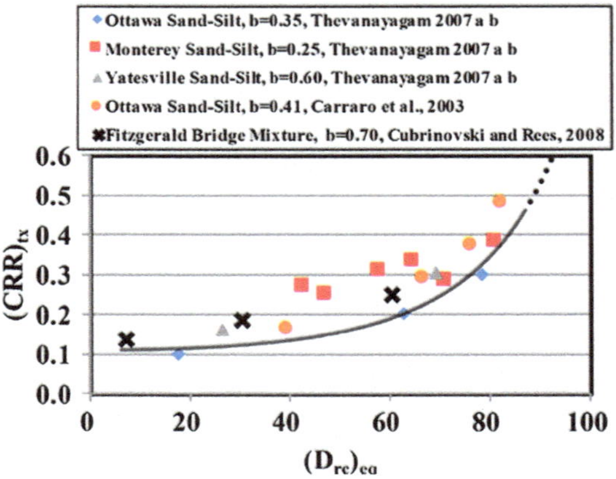

Fig. 6 Effect of silt content on CRR versus $(D_{\mathrm{rc}})_{\mathrm{eq}}$

two subset data, respectively. The cyclic strength data falls in a narrow band in each case, indicating the unifying influence of $(e_c)_{eq}$ and $(e_f)_{eq}$, respectively, compared to the data set in Fig. 4. This idea was tested for more than 30 silty sands and found to hold well (Veluchamy 2012). Such examination included data sets reported by Thevanayagam et al. (2002), Thevanayagam (2007a, b), Carraro et al. (2003), Cubrinovski and Rees (2008), and numerous others.

Figure 6 shows a relationship between CRR (at 15 cycles to liquefaction, obtained from cyclic triaxial tests) versus $(D_{\mathrm{rc}})_{\mathrm{eq}}$, for many different sands and sand–silt mixes. The data points for all soil mixes fall in a narrow band, further illustrating that the effects of silt content on CRR can also be characterized using the inter-grain contact density $(D_{\mathrm{rc}})_{\mathrm{eq}}$ of the soil. While the intergranular concept, implemented through a simple parameter b, does unify and narrow-down the spread of influence of silt particles on cyclic resistance, it does not completely capture the influence of other grain size/gradation characteristics. However, in general, measures of liquefaction resistance such as CRR, energy E_L (per unit volume of soil) required to cause liquefaction, shear modulus, and shear wave velocity correlate well with $(e_c)_{eq}$ or $(e_f)_{eq}$ and $(D_{\mathrm{rc}})_{\mathrm{eq}}$ (Thevanayagam 1999; Thevanayagam and Liang 2001; Kanagalingam and Thevanayagam 2006a, b; Thevanayagam 2007a, b).

2.3 *Effect of Silt Content and Pore Size on* k *and* c$_v$

For the same Ottawa sand–silt soil mixes discussed above in Fig. 4, laboratory tests were also conducted to measure k and c_v, based on stress and volume measurements made during consolidation of soil specimens prepared for monotonic and/or cyclic triaxial tests (Shenthan 2001; Thevanayagam et al. 2001).

Figure 7 shows the variation of k against silt content. k decreases steadily with an increase in silt content up to a threshold value of FC of about 25–30%. Beyond that FC, k did vary somewhat and was found to primarily depend on the inter-fine

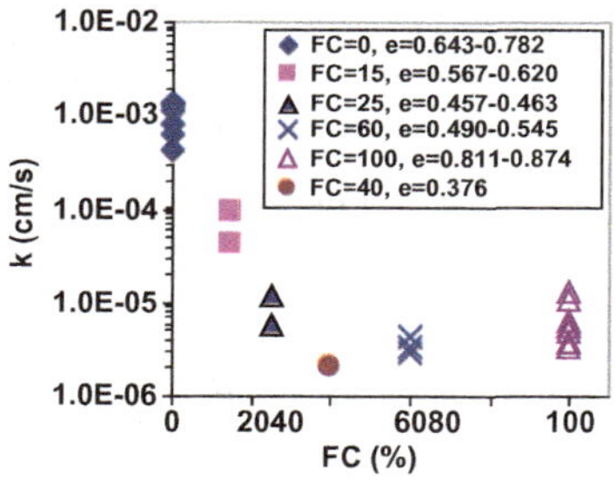

Fig. 7 k versus silt content

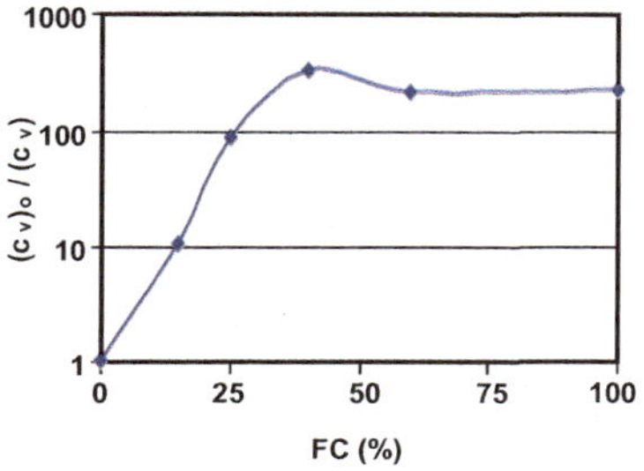

Fig. 8 c_{vo}/c_v versus silt content

void ratio. The initial reduction is considered to be due to the significant reduction in pore size with increase in silt content until the voids are fully filled by silt. Subsequent to the threshold FC, flow is primarily contained within the pores between the silt particles. In this regime, k did vary somewhat and was found to primarily depend on the inter-fine void ratio.

Figure 8 shows the normalized (c_{vo}/c_v) values at nearly the same $(e)_{eq}$ versus silt content, where c_v and c_{vo} are the coefficient of consolidation of Ottawa sand–silt mix and clean Ottawa sand, respectively. The direct and significant influence of silt content on k is reflected on c_v. (c_{vo}/c_v) increases (viz. c_v decreases) steadily with an increase in silt content up to a threshold value of silt content (FC_{th}) and it is little affected with further increase in silt content.

2.4 *Summary*

The above combined observations from 2.1 and 2.2 indicate that a sand and sand–silt mix show similar liquefaction resistance as the host sand or silt, when compared at the same contact density indices $(e_c)_{eq}$ or $(e_f)_{eq}$, or $(D_{rc})_{eq}$ respectively. Unlike the undrained cyclic resistance CRR which is primarily influenced by $(e)_{eq}$ or $(D_{rc})_{eq}$, c_v is primarily influenced by the significant effect of silt content on k. c_v has significant implication on consolidation behavior of a silty sand compared to that of a host sand at the same $(e_c)_{eq}$. This difference in c_v has a paramount influence on any potential correlation between cyclic resistance (which is primarily influenced by intergranular contact density) and cone resistance (which is influenced by intergranular contact density and c_v as well). This is discussed in detail in Sect. 3.

3 Effect of $(D_{rc})_{eq}$ and c_v on q_{c1N}

3.1 Model Cone Tests

Penetration of a CPT probe into a saturated soil causes highly non-uniform stresses, shear strains, and excess pore pressures in the soil around the probe. The rate of penetration, geometry of the penetrating probe, stress–strain characteristics of the soil, and its consolidation characteristics are expected to influence the rate of dissipation of the excess pore pressures and therefore the effective stress regime around the cone.

If the stress–strain characteristics of a sand and silty sand are the same, and the probe geometry is the same, one could expect the same penetration resistance if the penetration rate is so slow that it is essentially a drained penetration with no excess pore pressures. Likewise, if the penetration is extremely fast, essentially making it a full undrained penetration, one could expect the same penetration resistance for both soils. At any other rates of penetration, one would expect different pore pressure responses for sand and silty sand due to differences in consolidation characteristics. Therefore, different penetration resistances would be expected in sand and silty sand, even if they have the same stress–strain characteristics.

To test this hypothesis, a series of exploratory model cone tests were performed on clean Ottawa sand and silty sand mix at 25% silt content. Cone penetration tests were done on each soil mix while they were dry and again while they were fully saturated. The test hypotheses were that

(a) in the absence of excess pore pressures, cone resistance is affected by primarily intergranular contact density. Therefore, the cone resistances must be very similar for both soils at the same $(D_{rc})_{eq}$, when they are dry, and
(b) when the soil is saturated, cone resistance is influenced by intergranular contact density and k and c_v. Therefore, cone resistance of saturated sand and saturated silty sand must be different even when the intergranular contact density $(D_{rc})_{eq}$ is the same.

The tests were conducted using a scaled-model CPT probe of 1.27 cm diameter (d) pushed into soils contained in a cylindrical chamber (internally 50.8 cm dia, 49 cm high, Fig. 9). Dry specimens were prepared by dry pluviation and gentle compaction. Saturated specimens were prepared by dry pluviation followed by percolation of carbon dioxide and water saturation using back-pressure technique. Soil dry weight, initial moisture content (typically ranging in 0–0.3%), volume of the container when the soil is filled, soil volume changes during saturation and

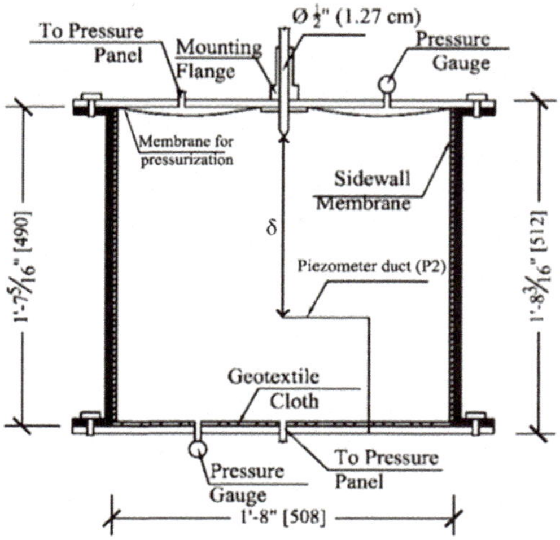

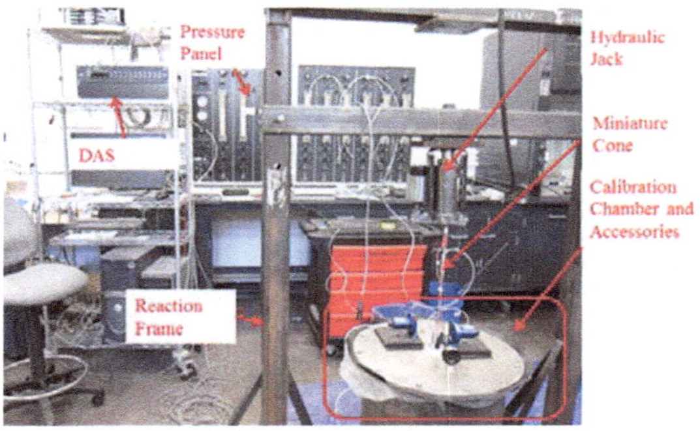

Fig. 9 **a** CPT chamber, **b** CPT chamber test system

consolidation, the volume of water introduced to the specimen, and compliance of the chamber were recorded. They were used to estimate the final void ratio of the specimen after consolidation, prior to cone penetration.

Two different methods were used to estimate the final void ratio e: (a) based on test chamber volume after correcting for settlement in soil height during consolidation, and small chamber expansion due to pressure, and (b) based on dry weight of sand and total volume of water in the saturated specimen, which does account for volume changes during saturation. The measured void ratio was used to calculate $(D_{rc})_{eq}$. Where applicable, this is shown by a band in Fig. 9a, introduced later.

The effective vertical stress in the soil prior to cone penetration was typically about 50 kPa. Sidewall soil friction in the chamber was addressed using a lubricated thin film placed between the soil and the side wall of the chamber. A needle-tip piezometer was planted at a vertical distance *d* below the cone tip (Fig. 9a) to monitor how pore pressure develops as cone is penetrated. Porewater pressure at the piezometer needle-tip and the distance d were continuously monitored during cone penetration until the cone tip reached the needle-piezometer-tip (viz. until $d = 0$). The measured cone resistance data was normalized for 100 kPa of confining stress to obtain q_{c1N}.

3.2 Model Cone Test Results

Figure 10a shows measured q_{c1N} versus $(D_{rc})_{eq}$ data from the above tests. OS-00 refers to Ottawa sand without any silt content. OS-25 refers to Ottawa sand with

Fig. 10 Effect of silt (k and c_v) on q_{c1N} and porewater pressure (Δu)

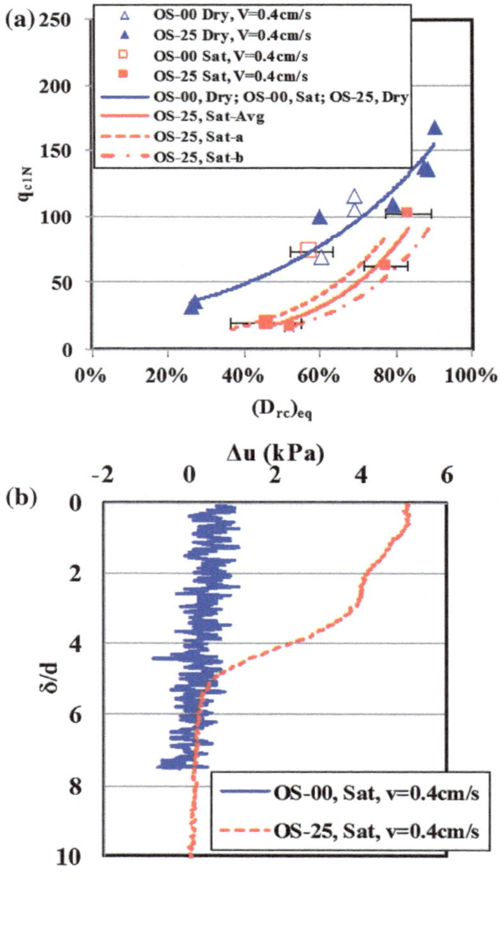

25% silt content. The soil mix is the same Ottawa sand–silt soil mix indicated in data sets shown in Figs. 4, 5, 6, 7, and 8. The following observations can be made.

- When the sand (open triangle) and silty sand (solid blue triangle) were dry, the q_{c1N} versus $(D_{rc})_{eq}$ data for sand and silty sand follow a similar trend, not indicating any significant effect of silt content on q_{c1N}. This is because the cone resistance in dry soils is influenced by the intergranular contact density, and not (air) permeability or (air) coefficient of consolidation. The tests were fully drained for both soils. The measured q_{c1N} is a predominantly mechanical response of the soil without any influence by permeability k, coefficient of consolidation c_v or pore pressures around the cone.
- When the q_{c1N} versus $(D_{rc})_{eq}$ data for dry sand (open triangle) is compared with saturated sand (open red square), again, no difference is observed in the q_{c1N} versus $(D_{rc})_{eq}$ trend. Moreover, no excess pore pressures were observed for saturated sand as shown in Fig. 10b (blue line). This q_{c1N} versus $(D_{rc})_{eq}$ trend is because, for all practical purposes, the cone penetration is almost fully drained response in saturated sand, and hence does not noticeably differ from q_{c1N} for dry sand at the same $(D_{rc})_{eq}$.
- Saturated silty sand (solid red square) shows significantly low q_{c1N} compared to q_{c1N} for saturated sand (open square) at the same $(D_{rc})_{eq}$. This reduction in q_{c1N} for saturated silty sand is because, the presence of silt reduces the k and c_v in saturated silty sand, and therefore induces noticeable pore pressures (Fig. 10b, red dash line), and makes the penetration process in saturated silty sand partially or fully undrained response, whereas the penetration in sand is fully or nearly fully drained response. Therefore, they have different q_{c1N} response at the same $(D_{rc})_{eq}$.
- Likewise, saturated silty sand (solid red squares) exhibits a low q_{c1N} compared to dry silty sand (solid triangle), at the same $(D_{rc})_{eq}$, for the same reasons as above.

Figure 10b shows the pore pressure at a normalized distance δ/d below the cone tip as the cone was continuously pushed into the soil in saturated silty sand (25% silt, red dash line) compared to saturated sand (blue line). A decreasing δ/d means the cone tip was approaching the piezometer location. $\delta/d = 0$ means that the cone tip has reached the piezometer needle-tip location and the test was terminated at that point.

There was no noticeable change in pore water pressure for saturated sand for the entire penetration process, indicating mostly fully drained cone penetration. In silty sand, no pore pressures were felt at the piezometer location, when the piezometer tip was far below the cone tip until the cone tip reached a normalized distance δ/d of about 4. From then on, there was noticeable change in porewater pressure increasing as the cone tip approached the piezometer location. This observation indicates that pore pressures are generated around the cone in a saturated silty sand and it is not fully dissipated immediately, unlike in saturated sand. Only a partial drainage occurs around the cone tip in saturated silty sands. These tests were performed using a small diameter cone (1.27 cm) and penetrated at a slow rate of 0.4 cm/s. A standard cone diameter is 4.37 cm penetrated at 2 cm/s. It would be expected that the porewater dissipation rates around the large standard cone would be much slower by about an order of magnitude compared to the laboratory model cone used in the above experiments, and therefore larger porewater pressures would remain around such a standard cone in saturated silty sands and influence the q_{c1N} compared to saturated sand.

These model cone tests are preliminary and the interpretations reported herein are subject to further research, and analysis, with several series of new data sets to be generated on this soil mix and other soil mixes, including tests involving larger size cones and faster penetration rates before definitive conclusions can be made.

3.3 Summary

Despite limitations of the data sets involved, when compared at the same $(D_{rc})_{eq}$, it is noted that the presence of silt, which influences c_v, q_{c1N} of saturated silty sand is lower than that of clean sand (Fig. 10a), whereas the cyclic resistances are nearly the same (Fig. 6) for saturated sand and silty sand. This points to a three-way CRR-q_{c1N}-T relationship. Such a rational relationship could lead to use of in situ q_{c1N} and in situ c_v to determine CRR to perform an improved liquefaction screening analysis in sands and non-plastic silty soils.

4 CRR-q_{c1n}-T for Liquefaction Screening

The observations in Sects. 2 and 3 indicate that (a) CRR is related to $(D_{rc})_{eq}$. and (b) q_{c1N} ia related to $(D_{rc})_{eq}$ and c_v. When the size of cone d and penetration velocity v are included, further analysis showed that q_{c1N} ia related to $(D_{rc})_{eq}$ and T. This indicates the fines-content-dependent liquefaction screening chart shown in Fig. 1 could be further refined to develop a more rational T-dependent CRR-q_{c1N}-T relationship (Thevanayagam et al. 2016a, b).

In an attempt to develop such a relationship, first the measured $(CRR)_{tx}$ versus $(D_{rc})_{eq}$ relationship, obtained from triaxial tests, shown in Fig. 6, was corrected for multi-shaking effects (Seed et al. 1978) to obtain an equivalent $(CRR)_{field}$ versus $(D_{rc})_{eq}$. Second, the model cone data shown in Fig. 10 and additional model cone data collected at different penetration rates (Sivaratnarajah 2016) were organized into a q_{c1N}-$(D_{rc})_{eq}$-T relationship. Based on these two sets of relationships, by matching the dat for the same $(D_{rc})_{eq}$, a CRR- q_{c1N} -T relationship was obtained as shown in Fig. 11.

Figure 11 shows the CRR-q_{c1N}-T relationship corresponding to each cone test, and the associated T values.

Fig. 11 Liquefaction screening: CRR-q_{c1N}-T

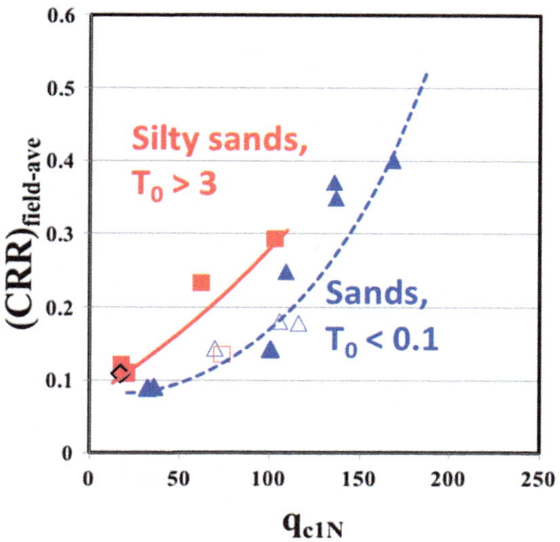

- The dashed blue line ($T < 0.1$) in Fig. 11 shows the relationship for saturated sand. The red line ($T > 3$) shows the relationship for saturated silty sand at 25% silt content.
- The T values for saturated sand (blue dash line) are less than 0.1 whereas the T values for the saturated silty sand (red line) are greater than 3. As T increases the CRR-q_{c1N} curve shifts to the left, for the same CRR, due to excess pore pressures developed around the cone in silty sands. For the same CRR, saturated silty sand (red line) shows a smaller q_{c1N} than saturated sand.

Figure 12 shows the CRR-q_{c1N}-T relationship (in Fig. 11) superimposed on the current CRR-q_{c1N}-FC liquefaction screening chart from Boulanger and Idriss (2014) (black lines for FC of 5, 15, 30 and 70%).

Fig. 12 Comparison: CRR-q_{c1N}-T and current screening methods

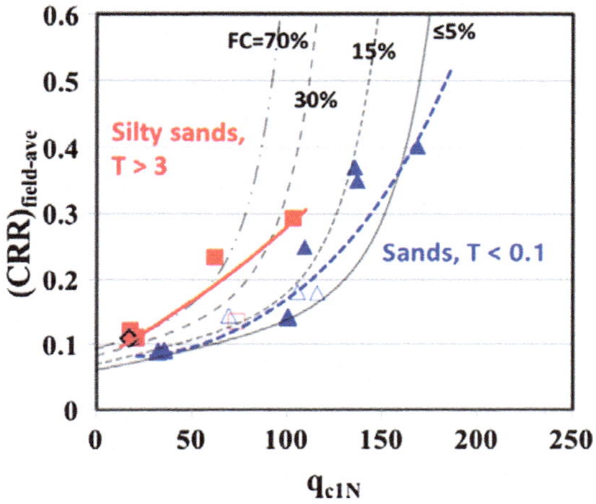

- The dash-blue line ($T < 0.1$) which corresponds to saturated sand follows the of B-I-2014 curve trend for sand.
- The red line ($T > 3$) that corresponds to saturated sand–silt mix (prepared at FC = 25%) follows the B-I-2014 trend for high silt content.
- This indicates that the different screening curves for different silt contents in screening charts (Fig. 1) are indeed manifestations of the effect of k and c_v (reflected in T), not just the gross amount of silt content.
- As the silt content increases, typically k and c_v decreases and T increases and hence q_{c1N} decreases for the same CRR. The reduction in q_{c1N} for silty sand is due to lack of rapid pore pressure dissipation around the cone in silty sand compared to the case for a clean sand at the same CRR.

5 Conclusions

Non-plastic silt content in silty sands affects intergrain contact density, liquefaction resistance CRR, permeability k, coefficient of consolidation c_v, and normalized cone penetration resistance q_{c1N}. Silt content affects the intergrain contact density of the silty sand.

When the effect of silt content on intergrain contact density is (approximately) accounted for, a sand and silty sand show similar CRR at the same equivalent void ratio $(D_{rc})_{eq}$.

Silt content also significantly affect k and c_v in silty soils compared to sand. Cone resistance is sensitive to $(D_{rc})_{eq}$ as well as c_v, cone diameter d and penetration velocity v. q_{c1N} of a silty sand is smaller than that of a sand at the same $(D_{rc})_{eq}$ when v and d are the same. This difference is apparently caused by different rates of drainage conditions that prevail around a cone tip in silty sand compared to that of sand. When velocity v, cone diameter d, and c_v are taken into account, there appears to exist a relationship between q_{c1N}, $(D_{rc})_{eq}$, and T.

Combining CRR versus $(D_{rc})_{eq}$ and q_{c1N}-$(D_{rc})_{eq}$-T relationships, as evident from Fig. 11, there exists a relationship between CRR-q_{c1N}-T. It appears what separates the CRR-q_{c1N} relationship (Fig. 1) for silty sands from sand are the effects of silt content on CRR, and the additional effects of silt content (namely permeability and consolidation characteristics) on q_{c1N}.

A rational liquefaction screening relationship between CRR-q_{c1N}-T is likely. Such a screening method could be implemented in the field measurements q_{c1N} and piezocone measurements of c_v by stopping the cone penetration for just a few seconds when the cone tip reads notably high excess pore pressures indicating presence of silty sands, excluding clayey zones. Such a rational CRR-q_{c1N}-T liquefaction screening method could help reduce the level of uncertainty in the current screening methods and advance the state of practice in liquefaction triggering analysis.

Acknowledgements Financial support for this research was provided by National Science Foundation, and USGS NEHRP program award no. 07HQGR0113, and MCEER Highway Project 094, sponsored by the FHWA. This assistance is greatly appreciated. T. Shenthan, R. Nashed, N. Ecemis, Y. Liang, T. Kanagalingam, and M. Reza are thanked for their assistance in this work.

References[1]

Andrus, R.D., and K.H. Stokoe. 2000. Liquefaction resistance of soils from shear-wave velocity. *Journal of Geotechnical and Geoenvironmental Engineering, ASCE* 126 (11): 1015–1025.

Boulanger, R.W., and I.-M. Idriss. 2014. *CPT and SPT based liquefaction triggering procedure*, 138 p. Report No. UCD/CGM-14-01. Davis: University of California.

Carraro, J.A.H., P. Bandini, and R. Salgado. 2003. Liquefaction resistance of clean and nonplastic silty sands based on cone penetration resistance. *Journal of Geotechnical and Geoenvironmental Engineering* 129 (11): 965–976.

Cubrinovski, M., and S. Rees. 2008. *Effects of fines on undrained behavior of sands*. University of Canterbury Research Repository.

Idriss, I.M., and R.W. Boulanger. 2008. *Soil liquefaction during earthquakes*, 261 p. Monograph MNO-12, EERI, Oakland, CA.

Kanagalingam, T., and S. Thevanayagam. 2006. *Contribution of fines to the compressive strength of mixed soils*. Discussion, Geotechnique.

Kanagalingam, T., and S. Thevanayagam. 2006. Energy dissipation and liquefaction assessment in sands and silty soils. In *GeoCongress 2006: Geotechnical engineering in the information technology age*.

Koester, J.P. 1994. The influence of fines type and content on cyclic resistance. In *Ground failures under seismic conditions*, 17–33. Geotech Spec. Publ. No. 44. New York: ASCE.

NCEER. 1997. In *Proceedings of the NCEER workshop on evaluation of liquefaction resistance of soils*, 276 p. NCEER Technical Report NCEER-97-0022, prepared by Youd, T.L, and I.M. Idriss.

Ni, Q., T.S. Tan, G.R. Dasari, and D.W. Hight. 2004. Contribution of fines to the compressive strength of mixed soils. *Geotechnique* 54 (9): 561–569.

Polito, C.P., and J.R., Martin II. 2001. Effects of nonplastic fines on the liquefaction resistance of sands. *Journal of Geotechnical and Geoenvironmental Engineering ASCE* 127 (5): 408–415.

Rahman, M.M., S.R. Lo, and C.T. Gnanendran. 2008. On equivalent granular void ratio and steady state behaviour of loose sand with fines. *Canadian Geotechnical Journal* 45 (10): 1439–1456.

Robertson, P.K., and C.E. Wride. 1998. Evaluation of cyclic liquefaction potential using the cone penetration test. *Canadian Geotechnical Journal* 35 (3): 442–459.

Schneider, J.A., B.M. Lehane, and F. Schnaid. 2007. Velocity effects on piezocone tests in normally and overconsolidated clays. *International Journal of Physical Modelling in Geotechnics* 7 (2): 23–34.

Seed, H. B., G. R. Martin, and C. K. Pyke. 1978. Effects of multi-directional shaking on pore pressure development in sands. *Journal of the Geotechnical Engineering Division, ASCE* 104 (1): 27–44.

Seed, H.B, I.M. Idriss, and I. Arango. 1983. Evaluation of liquefaction potential using field performance data. *Journal of the Geotechnical Engineering, ASCE* 109 (3): 458–482.

Shenthan, T. 2001. *Factors affecting liquefaction mitigation in silty soils using stone columns*. M.S. thesis, State University of New York at Buffalo, NY.

[1] A large number of literature on the effects of silt content on cyclic resistance of soils is not explicitly listed herein, for brevity. They are summarized in Veluchamy (2012).

Sivaratnarajah, U. 2016. *Effect of non-plastic silt on cone resistance*. M.S. thesis, State University of New York at Buffalo, NY.

Thevanayagam, S. 1998. Effects of fines and confining stress on undrained shear strength of silty sands. *Journal of Geotechnical and Geoenvironmental Engineering* 124 (6): 479–491.

Thevanayagam S. 1999. Liquefaction and shear wave velocity characteristics of silty/gravely soils—implications for bridge foundations. In Proceedings of 15th US–Japan Bridge Work-shop, PWRI, Japan, 133–47, ed. K. Nishikawa.

Thevanayagam, S., and S. Mohan. 2000. Intergranular state variables and stress-strain behaviour of silty sands. *Geotechnique* 50 (1), 1–23.

Thevanayagam, S., G.R. Martin, T. Shenthan, and J. Liang. 2001. Post-liquefaction pore pressure dissipation and densification in silty soils. In *Proceedings of 4th international conference on recent advances in geotechnological earthquake engineering & soil dynamics*, ed. S. Prakash, Paper# 4.28.

Thevanayagam, S., and Liang, J. 2001. Shear wave velocity relations for silty and gravely soils. In *Proceedings of 4th international conference on soil dynamics & earthquake engineering*, San Diego.

Thevanayagam, S., T. Shenthan, S. Mohan, and J. Liang. 2002. Undrained fragility of sands, silty sands and silt. *Journal of Geotechnical and Geoenvironmental Engineering ASCE* 128 (10): 849–859.

Thevanayagam, S., and G.R. Martin. 2002. Liquefaction in silty soils—screening and remediation issues. *Soil Dynamics and Earthquake Engineering* 22: 1034–1042.

Thevanayagam, S., N. Ecemis, T. Kanagalingam, and G.R. Martin. 2006. Liquefaction screening of silty soils using Cone penetration tests. In *Proceedings of 8th national conference on earthquake engineering,* San Francisco, CA.

Thevanayagam, S. 2007a. Intergrain contact density indices for granular mixes–I: Framework. *Journal of Earthquake Engineering and Engineering Vibration* 6 (2): 123–134.

Thevanayagam, S. 2007b. Intergrain contact density indices for granular mixes–II: Liquefaction resistance. *Journal of Earthquake Engineering and Engineering Vibration* 6 (2): 135–146.

Thevanayagam, S., U. Sivaratnarajah, and Q. Huang. 2016. A rational CPT-based liquefaction screening method—effect of silt content'—Invited lecture. In *Proceedings of 1st international symposium on soil dynamics and geotechnological sustainability*, 60–63, ed. Wang, G., G. Zhang, and D. Huang. HKUST, Hog Kong. ISBN 978-988-14032-4-7.

Thevanayagam, S., V. Veluchamy, Q. Huang, and U. Sivaratnarajah. 2016. Silty sand liquefaction, screening, and remediation. *Journal of Soil Dynamics and Earthquake Engineering* (in press).

Veluchamy, V. 2012. *Effect of sand/silt gradation on the undrained strength of silty sands.* Report submitted for the fulfillment of MS degree, University at Buffalo, NY.

Youd, T.L., I.M. Idriss, R.D. Andrus, I. Arango, G. Castro, J.T. Christian, R. Dobry, W.D.L. Finn, L.F. Harder Jr, M.E. Hynes, K. Ishihara, J.P. Koester, S.S.C. Liao, W.F. Marcuson III, G.R. Martin, J.K. Mitchell, Y. Moriwaki, M.S. Power, P.K. Robertson, R.B. Seed, and K.H. Stokoe II. 2001. Liquefaction resistance of soils: summary report from the 1996 NCEER and 1998 NCEER/NSF workshops on evaluation of liquefaction resistance of soils. *Journal of Geotechnical and Geoenvironmental Engineering, ASCE* 127 (10): 817–833.

Zlatovic, S., and K. Ishihara. 1997. Normalized behavior of very loose non-plastic soils: effects of fabric. *Soils and Foundations* 37 (4): 47–56.

Insight into the Role of Osmotic Suction in Soil Behavior

Sudhakar M. Rao

1 Introduction

Understanding the engineering response of clays to changes in pore salt concentration is relevant in situations; such as, infiltration of chemical solutions in clay liners (Barbour and Yang 1993), chemical spillages during industrial operations (Rao and Sridharan 1993), sea-water intrusion in coastal aquifers from increased sea levels (Carretero et al. 2013) or lowering of groundwater table from excessive pumping (Pulido-Leboeuf 2004; Melloul and Goldenberg 1997). Alterations in pore water salinity and associated ion exchange reactions influence electrochemical stress (*A*–*R*) and differential osmotic stress ($\Delta\pi$) of clay-water suspensions (Barbour and Fredlund 1989; Di Maio 1996; Rao and Shivananda 2005; Rao et al. 2006); differential osmotic stress ($\Delta\pi$) arise from differences in dissolved salt concentration in pore water and external reservoir and is calculated as: where, R is universal gas constant (8.32 L kPa/mol K), T is absolute temperature (°K), and M_1 and M_2 are dissolved salt concentrations in pore-water (M_1) and consolidometer reservoir water (M_2) respectively. The electrochemical stress (*A*–*R*) is contributed by van der Waals (*A*) attraction and diffuse ion layer repulsion (*R*) (Barbour and Fredlund 1989; Mitchell and Soga 2005). The work presented below is from published research of the author and are appropriately quoted.

S. M. Rao (✉)
Department of Civil Engineering, Indian Institute of Science, Bengaluru 560092, India
e-mail: msrao@civil.iisc.ernet.in

K. Ilamparuthi and R. G. Robinson (eds.), *Geotechnical Design and Practice*,
Developments in Geotechnical Engineering,
https://doi.org/10.1007/978-981-13-0505-4_3

1.1 Influence of Osmotic Suction Dissipation on Volumetric Strain (Rao et al. 2006; Rao and Thyagaraj 2007a)

Expansive clay specimens were compacted at 28% water content and dry density of 1.42 Mg/m^3; the degree of saturation and matric suction of specimens corresponded to 84% and 2000 kPa respectively. The specimens were inundated with 0.1–4 M (molar) or 0.1–4 N (normal; normality and molarity is identical for sodium chloride solutions) sodium chloride solutions in conventional oedometer cells at net vertical stress of 6.25 kPa. The dissolved salts concentrations in pore water of the compacted clay specimens (0.3 g/L) were smaller than concentrations of reservoir solutions (5.84–233.6 g/L, NaCl solutions) and induced differential osmotic suction stress of 391–8968 kPa in pore water. The matric suction and osmotic suction gradient in the compacted clay dissipated through inward flow of salt solution and outward flow of constituent H_2O molecules of pore water. The outward flow of constituent H_2O molecules of pore water reduced pore water pressure, increase average skeleton stress with void ratio reduction (Barbour and Fredlund 1989). In contrast, the inward flow of salt solution promoted swelling of the compacted expansive clay specimens, as the external load (6.2 kPa) was lower than the swell pressure of the clays (56–160 kPa).

Figure 1 plots axial deformations with time for specimens inundated with distilled water, 0.1, 0.4, 1, and 4 N NaCl solutions respectively. The compacted clay specimens swell on inundation with NaCl solutions, which implies that dissipation of matric suction and the attendant growth of diffuse ion layer repulsion dominates

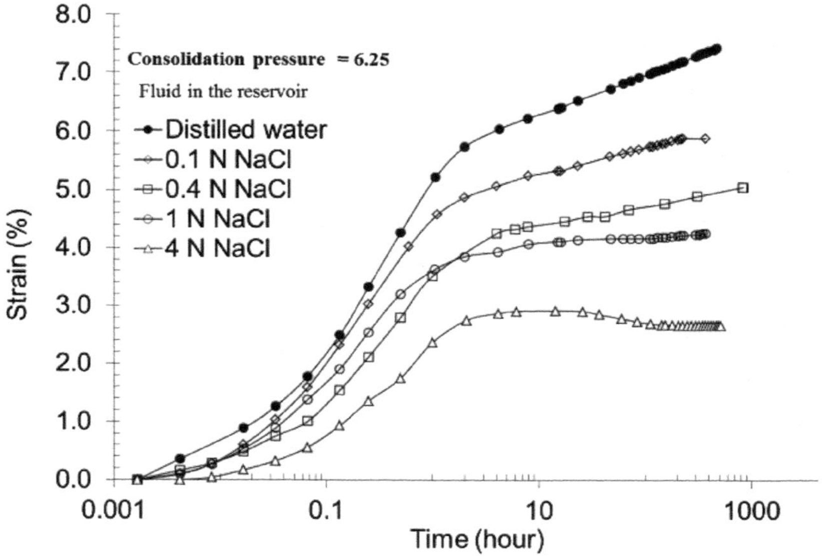

Fig. 1 Time-swelling behavior of expansive clays

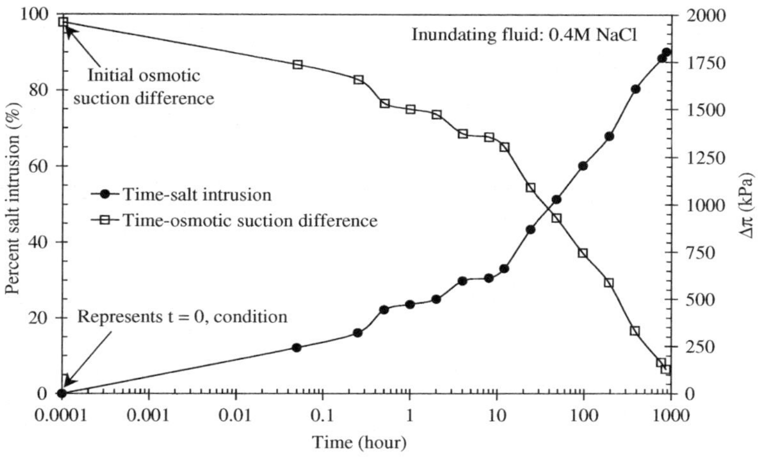

Fig. 2 Dissipation of osmotic suction with time of expansive clays

compacted clay behavior at all salt concentrations (van Olphen 1963; Yong and Warkentin 1975).

Figure 2 traces osmotic suction difference ($\Delta\pi$ from Eq. 1) and percent swell as a function of time for compacted clays inundated with 0.4 M sodium chloride solution. Percent salt intrusion at time interval t is calculated as

$$\frac{\text{Salt in soil water extract at time } t}{\text{Salt in oedometer reservoir at } t = 0} \times 100\% \tag{1}$$

Plots in Fig. 2 reveal that 90% of salt intrusion and 93% dissipation of initial osmotic suction difference occur after 40 days of contact between clay and salt solutions. In comparison, Fig. 3 shows that large fraction of matric suction dissipates rapidly (80–100 min) and before mobilization of complete swells potential. The inward salt intrusion in response to dissipation of matric suction and osmotic

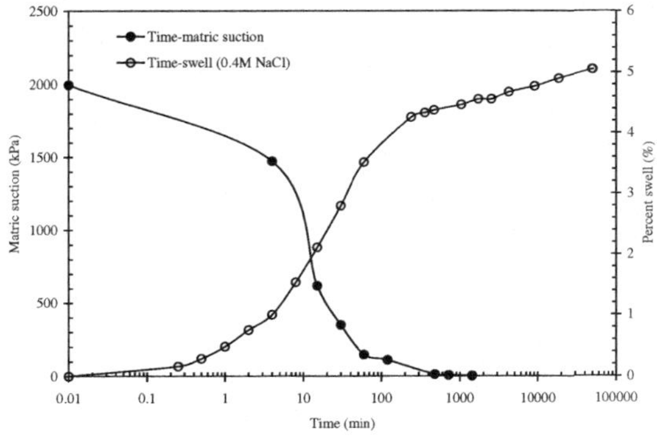

Fig. 3 Dissipation of matric suction and growth of swell with time of expansive clay

suction difference suppresses the diffuse ion layer thickness around the clay particles and lowers the swell magnitude of compacted clay specimens inundated with 0.4–4 M sodium chloride solutions in reference to specimen inundated with distilled water (Fig. 1).

1.2 *Role of Osmotic Suction in Swell Mechanism (Rao et al. 2013)*

Swelling of unsaturated expansive clays on contact with water occurs in two stages. The first stage pertains to inter-layer separations between 10 and 22 Å and is driven by moisture absorption on clay surfaces (crystalline swelling). The second swelling occurs at inter-layer separations beyond 22 Å and is associated with double-layer/osmotic swelling. The inter-layer expansion due to adsorption of water molecules in the crystalline swelling region (10–22 Å) is function of layer charge, exchangeable cation and attraction between water molecules and the polar surface groups of the clay mineral (van Olphen 1963; Yong 1999, Saiyouri et al. 2004; Villar 2007). Swelling beyond 22 Å can be calculated as an osmotic pressure using Van't Hoff's equation:

$$\pi = \mathrm{RT}(C_\mathrm{c} - 2C_0), \tag{2}$$

where π represents the osmotic or swelling pressure, R is the gas constant, T is the absolute temperature, C_c is concentration of cations at midpoint of two interacting clay platelets in the micro-structural space (C_c in mol/L) and C_0 is the cation concentration (mol/L) in the bulk solution in macro-pores. Increase in C_0 would reduce π and the contribution of double-layer repulsion to swell potential. The osmotic pressure is identical with the double-layer repulsion energy of two plates as computed by the equation

$$p = 2nkT(\cos\mathrm{h}\, u - 1), \tag{3}$$

where p is the swell pressure, n is the ion concentration in the bulk solution, k is Boltzmann constant, T is absolute temperature and u is the mid-plane potential between two parallel clay plates (van Olphen 1963; Yong and Warkentin 1975; Mitchell and Soga 2005; Karnland 1997). The experimental data pertains to oedometer tests with compacted expansive soil specimens described in Sect. 1.2. The concentration of cations at midpoint of two interacting clay plates in the inter-lamellar space (C_c in mol/L) is calculated as (Yong and Warkentin 1975; Karnland 1997)

$$C_\mathrm{c} = \frac{\pi^2}{z^2 B (d + x_0)^2 10^{-16}}, \tag{4}$$

where z = valence of exchangeable cation, d is the half distance between two clay plates in Å (10^{-10} m), x_0 is a correction factor (1–4 × 10^{-8} cm) and B is a constant with the value of 1 × 10^{18} cm/mol and represents the combined effects of temperature and dielectric constant (values from Yong and Warkentin 1975; Karnland 1997). Inserting appropriate values for d (8 × 10^{-8} cm), x_0 (2 × 10^{-8} cm) and z (1.99-weighted average value for the clay specimen) yields C_c value of 0.25 M (molar) for the compacted clay specimen used in this study. The compacted clay specimens were wetted with distilled water, 0.1, 0.4, 1 and 4 M sodium chloride solutions in the oedometer tests at net vertical stress of 6.25 kPa. It is assumed that wetting with salt solutions rendered C_0 equal to concentration of wetting fluids (0.1–4 M sodium chloride solution). Comparison of the C_c and C_0 values lead to the following observations. The C_c exceeds the C_0 values (solute concentration in bulk solution) when compacted clay specimens are wetted with distilled water (C_0 of clay specimen wetted with distilled water corresponds to 0.005 M NaCl) and 0.1 M sodium chloride solution in the oedometer tests. The C_c values are however smaller than the C_0 values for clay specimens wetted with 0.4, 1 and 4 M sodium chloride solutions in the oedometer tests. Figure 1 had presented the percent swell versus time plots of compacted clay specimens wetted at net vertical stress of 6.25 kPa in the oedometer tests. The compacted clay specimens wetted with distilled water, 0.1, 0.4, 1, and 4 M NaCl solutions develop swell potentials of 7.4, 5.9, 5.1, 4.2, and 2.7% respectively. Equation (2) states that osmotic swell occurs when $C_c > C_0$ condition is met. This condition is met, when the compacted specimens are inundated with 0.1 M sodium chloride solution and are not satisfied upon inundation with the relatively more saline, 0.4, 1, and 4 M sodium chloride solutions. Wetting sof the compacted clay specimens with saline solutions restricts the growth of diffuse ion layers. Owing to the restriction in diffuse ion layer growth, the swell potentials of specimens that do not meet $C_c < C_0$ condition are satisfied by crystalline swelling alone. Comparatively, compacted clay specimens inundated with relatively less saline solutions (0.005–0.1 M sodium chloride) need both crystalline and osmotic/double-layer swelling to satiate the swell potential.

1.3 Equivalent Net Stress from Osmotic Suction (Rao and Thyagaraj 2007b)

Rao and Thyagaraj (2007b) have indicated that osmotic flow of salt solution from external reservoir to pore water of the clay induces an additional equivalent net stress component (p_π)

$$p_\pi = \alpha\Delta\pi = \alpha\left(\pi_{1f} - \pi_{2f}\right) \tag{5}$$

In Eq. (5), osmotic efficiency (α) varies between 0 and 1 and determines the magnitude of $\Delta\pi$ contribution to true mean effective stress. Values of $\alpha = 0$ and 1, imply imperfect and perfect semi-permeable membrane behavior respectively. In Eq. (5), π_{1f} represents equilibrium osmotic suction in pore water of clay and π_{2f} represents equilibrium osmotic suction of the external (reservoir) solution. When ions flow from external reservoir to pore-water in response to differential osmotic stress, it induces an additional equivalent net stress (p_π) and enhances the true effective stress.

1.4 Calculation of Osmotic Efficiency (Sects. 2.4–2.7 from Rao et al. 2016)

The osmotic efficiency (α) can be calculated from the Fritz-Marine Membrane Model (Fritz 1986; Keijzer 2000; Maine and Fritz 1981); the method is illustrated for kaolinites slurries prepared with 0.02 and 0.4 M sodium chloride solutions and subjected to 1-dimensional consolidation in conventional oedometers (Rao et al. 2016). The model is based on the premise that at osmotic equilibrium, thermodynamic forces acting across the clay membrane are balanced by the sum of mechanical-frictional forces between ions and water/membrane. Anions attempting to migrate into the pores are repelled by the negative mid-plane potential; the tendency to exclude anions also hinders cation migration because cations need anions to maintain electrical neutrality. Consequently, as the porosity of clay decreases, increased overlap of diffuse ion layers enhances frictional resistance between anions and clay membrane that in turn enhances the osmotic efficiency of the clay. The model calculates osmotic efficiency (α) as

$$\alpha = 1 - \frac{K_s(R_w + 1)}{\left[\left(R_w \frac{\bar{C}_a}{\bar{C}_c} + 1\right) + R_{wm}\left(R_m \frac{\bar{C}_a}{\bar{C}_c} + 1\right)\right] n} \tag{6}$$

In Eq. (6), $K_s = C_a/C_s$, where C_a is the anion concentration in pore water (mol/cm^3) and C_s represents average solute concentration (mol/cm^3) of pore water + reservoir water. R_w represents the frictional resistance between solutes and pore water and equals 1.63 for the NaCl amended kaolinite specimens used in the study (Fritz and Maine 1983). R_m represents the tendency of ions to be retarded by frictional resistance with membrane walls. Neglecting electrostatic effects, R_m is obtained by considering the hydrated radii of ions and ranges between 1.13 and 1.80 for NaCl–H_2O system (Fritz 1986; Keijzer 2000; Kharaka and Berry 1973; Marcus 1997). R_{wm} represents the friction of anions with both membrane (m) and pore water (w) and is >1 for compact clay ($n \ll 40\%$) and reduces with increasing porosity (Keijzer 2000). As porosity tends to zero, the contribution from R_{wm} can transform even, low to non-active clays, such as illite and kaolinite into ideal membranes (osmotic efficiency unity).

As porosity tends to zero, the contribution from R_{wm} can transform even, low to non-active clays, such as illite and kaolinite into ideal membranes (osmotic efficiency unity).

1.5 Incorporation of p_π *in True Effective Stress*

Mitchell and Soga (2005) define true inter-granular stress as:

$$\sigma_1' - \sigma' = A - R \tag{7}$$

where σ'_1 is true inter-granular stress and is contributed by σ' (conventional effective stress) and electrochemical stress (*A–R*). The true inter-granular stress (Budhu 2007; Mitchell and Soga 2005) is given as

$$\sigma' = \sigma - u, \tag{8}$$

where σ is the total stress and u is the pore water pressure. According to Fredlund et al. (2012) the osmotic stress (p_π) is an independent, isotropic, stress state variable. The physico-chemical contributions to true effective stress (σ'_1) are modified to incorporate p_π component (in addition to *A–R*); the true effective stress (σ'_1) of Mitchell and Soga (2005) becomes

$$\sigma_1' = \sigma' + (A - R) + p_\pi \tag{9}$$

Table 1 presents the osmotic efficiency calculated from the Marine and Fritz model (1981, Eq. 6) for the Na-kaolinite specimens at different consolidation pressures. R_w, the frictional resistance between ions (cations and anions) and pore water is assigned value of 1.63 (Fritz 1986). R_m, the friction of anions and cation with clay membrane structure is assigned value of 1.8 (Fritz 1986). R_{wm} term represents the frictional resistance of anions with both membrane and pore water and is strong function of porosity.

At given porosity (n), the least R_{wm} that yielded positive α (all other parameters remaining constant) defined the R_{wm} of the specimen at given consolidation pressure (Table 1). The porosity (n) values were obtained from experimental void ratios at different consolidation pressures (Table 1). C_a and C_c represent experimental chloride and sodium ion concentrations in pore water (mol/cm^3) at given consolidation pressure, while, C_s represents the average of $M_{1f} + M_{2f}$ (mol/cm^3) at each consolidation pressure (Table 1). Inserting the appropriate values gave the osmotic efficiency of the kaolinite specimens at different consolidation pressures (Table 1). The osmotic efficiencies of 0.4 M NaCl specimens (Table 1) range between 0.0049 and 0.0153 (n = 0.591–0.484) and between 0.005 and 0.1437 in case of the 0.02 M NaCl specimens (n = 0.567–0.439).

Table 1 Calculations of osmotic efficiency (α)

Consolidation pressure (kPa)	n (porosity)	C_a (mol/cm^3)	C_c (mol/cm^3)	C_s (mol/cm^3)	R_{wm}	α
0.4 M series						
6.25	0.591	0.00038	0.000397	0.000515	0.27	0.0049
25	0.569	0.00037	0.00038	0.000513	0.28	0.0061
100	0.537	0.000366	0.000373	0.0005095	0.34	0.0062
200	0.514	0.0003385	0.000343	0.0004945	0.34	0.015
400	0.484	0.000374	0.000376	0.0005215	0.48	0.0153
0.02 M series						
6.25	0.567	0.00002	0.0000185	0.0000305	0.1	0.005
25	0.539	0.0000189	0.0000176	0.0000305	0.1	0.0069
100	0.5	0.00001979	0.00001977	0.000036	0.1	0.044
200	0.474	0.0000205	0.0000186	0.0000395	0.1	0.103
400	0.439	0.0000181	0.0000175	0.000033	0.4	0.1437

Table 2 Physico-chemical parameters and true effective stress of kaolinites

Consolidation pressure (kPa)	e	$1/k$ (Å)	A (kPa)	R (kPa)	A–R (kPa)	π_{1f}–π_{2f} (kPa)	p_π (kPa)	σ_1' (kPa)
0.4 M series								
6.25	1.45	4.80	0.027	0	0.027	1466	7.31	13.6
25	1.32	4.90	0.029	0	0.029	1321	8.06	33.1
100	1.16	4.96	0.036	0	0.036	1257	7.79	107.8
200	1.06	5.17	0.039	0	0.039	1038	15.58	215.6
400	0.94	4.94	0.049	0	0.049	1258	19.25	419.3
0.02 M series								
6.25	1.31	22.27	0.021	0	0.021	82	0.41	6.7
25	1.17	22.84	0.028	0	0.028	67	0.46	25.5
100	1.00	21.55	0.030	0	0.030	64	2.80	102.8
200	0.90	22.21	0.038	0	0.038	37	3.80	203.8
400	0.785	22.90	0.040	0	0.040	54	7.77	407.8

The osmotic efficiencies of NaCl specimens increase with reduction in porosity apparently from increased interference to chloride migration into pores by the kaolinite micro-structure. Further, the 0.02 M specimen exhibits larger α than the 0.4 M specimen at given porosity possibly as the larger diffuse ion layer thickness (21.55–22.9 Å, Table 2) associated with these clay particles retard the passage of chloride ions into the pore space more effectively. The diffuse ion layer thickness ($1/\kappa$) of the kaolinite specimens at given consolidation pressure is obtained from the equation:

$$\frac{1}{\kappa} = \sqrt{\left(\frac{\varepsilon kT}{8\pi n' e'^2 v'^2}\right)}, \tag{10}$$

where ε is the dielectric constant of medium (water = 80), k is the Boltzmann's constant (1.38×10^{-16} erg/K), T is temperature in K (290 K), n' is sodium ion concentration in pore water at given consolidation pressure, e' is elementary charge (4.77×10^{-19} esu) and v' is sodium ion valence (unity).

1.6 Calculation of p_π

The contribution of p_π to true effective stress ($\sigma_1^!$) at various consolidation pressures are presented in Table 2. The differential osmotic stress [$\Delta\pi = (\pi_{1f} - \pi_{2f})$] range from 1038 to 1446 kPa for 0.4 M specimens (Table 2); multiplication of the differential osmotic stress with corresponding osmotic efficiency give p_π that range between 7 and 19 kPa for the 0.4 M specimens and from 0.41 to 7.77 kPa for 0.02 M specimens (Table 2). The attraction stress ranges from 0.027 to 0.049 kPa for the 0.4 M specimens from 0.021 to 0.04 kPa for 0.02 M specimens; both specimens are devoid of inter-particle repulsion (Table 2).

1.7 Void Ratio—True Effective Stress Plots

Figure 3 compares void ratio—true effective stress (σ'_1) plots of 0.02 and 0.4 M specimens. The 0.4 M specimens are less compressible than the 0.02 M specimens and exhibit larger void ratio at given true effective stress (Fig. 4). The 0.4 and 0.02 M series specimens are subjected to similar effective vertical stress (σ'_v: 6.25–400 kPa)

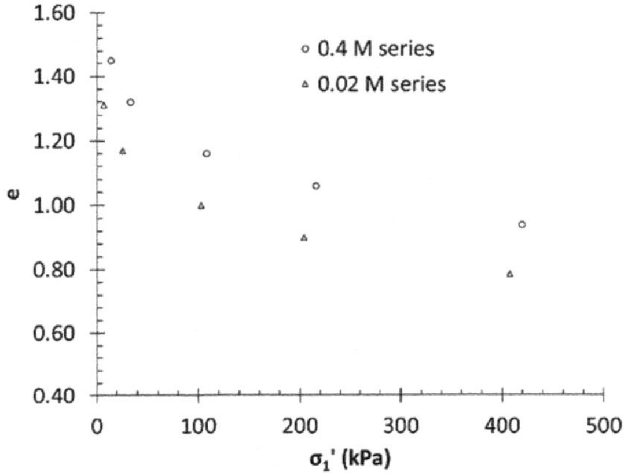

Fig. 4 Void ratio true effective stress plots of kaolinite

during consolidation; however of the two physico-chemical components, only p_π finitely contributes to true effective stress (Table 2). Further, the availability of higher p_π contribution (Table 2) enables the 0.4 M kaolinite specimens to sustain larger void ratio during consolidation (Fig. 4; Table 2).

2 Conclusions

Osmotic suction arises in compacted clays from differences in salt concentrations in pore water and external reservoir. Inflow of salt solution into expansive clay pores, reduces the swell potential. Flow of ion rich solution from external reservoir to pore-water in response to differential osmotic stress, induces an additional equivalent net stress (p_π) that enhances the true effective stress and lowers the swell potential. With non-swelling kaolinites, similar increase in true effective stress renders the clay less compressible. Contribution to swell from diffuse double-layer growth occurs only if the salt concentration in macro-pore solution is smaller than exchangeable cation concentration in the mid-plane region. Inundation of expansive clays with concentrated salt solutions favors crystalline swelling from moisture absorption by exchangeable cations and clay surface.

Contribution of additional equivalent net stress (p_π) to true effective stress is function of osmotic efficiency of the clay. The osmotic efficiencies of clay specimens increase with reduction in porosity from increased interference to anion migration into pores by the clay micro-structure. Further, clay specimens associated with thicker diffuse ion layers retard the passage of anions into pore space more effectively and exhibit larger osmotic efficiencies expansive clays with concentrated salt solutions favors crystalline swelling from moisture absorption by exchangeable cations and clay surface.

References

Barbour, S.L., and D.G. Fredlund. 1989. Mechanics of osmotic flow and volume change in clay soils. *Canadian Geotechnical Journal* 26: 551–562.

Barbour, S.L., and N. Yang. 1993. A review of the interaction of clay-brine interactions on the geotechnical properties of Ca-montmorillonitic clayey soils from Western Canada. *Canadian Geotechnical Journal* 30: 920–934.

Budhu, M. 2007. *Soil mechanics and foundations*. New York: Wiley.

Carretero, S., J. Rapaglia, H. Bokuniewicz, and E. Kruse. 2013. Impact of sea-level rise on salt water intrusion length into the coastal aquifer, Partido de La Costa, Argentina. *Continental Shelf Research* 61–62: 62–70.

Di Maio, C. 1996. Exposure of bentonite to salt solution: Osmotic and mechanical effects. *Geotechnique* 46: 695–707.

Fredlund, D.G., H. Rahardjo, and M.D. Fredlund. 2012. *Unsaturated soil mechanics in engineering practice*. New York: Wiley.

Fritz, S.J. 1986. Ideality of clay membranes in osmotic process: A review. *Clay Clay Minerals* 34: 214–223.
Fritz, S.J., and I.W. Marine. 1983. Experimental support for a predictive osmotic model of clay membranes. *Geochimica et Cosmochimica Acta* 47: 1515–1522.
Karnland, O. 1997. Bentonite swelling pressure in strong NaCl solutions—Correlations between model calculations and experimentally determined data. *SKB technical report* 97–31, Swedish Nuclear Fuel and Waste Management Co, Stockholm, Sweden.
Keijzer, Th. 2000. *Chemical osmosis in natural clayey materials*. Geologica Ultraiectina 196, PhD thesis, Utrecht University.
Kharaka, Y.K., and F.A.F. Berry. 1973. Simultaneous flows of water and solutes through geological membranes: I. Experimental investigation. *Geochimica et Cosmochimica Acta* 37: 2577–2603.
Marcus, Y. 1997. *Ion properties*. New York: Marcel Dekker.
Marine, I.W., and S.J. Fritz. 1981. Osmotic model to explain anomalous hydraulic heads. *Water Resources Research* 17: 73–82.
Melloul, A.J., and L.C. Goldenberg. 1997. Monitoring of seawater intrusion in costal aquifers: Basics and local concerns. *Journal of Environmental Management* 51: 73–86.
Mitchell, J.K., and K. Soga. 2005. *Fundamentals of soil behavior*, 3rd ed. Hoboken, NJ: Wiley.
Pulido-Leboeuf, P. 2004. Seawater intrusion and associated processes in a small coastal complex aquifer (Castell de Ferro, Spain). *Applied Geochemistry* 19: 1517–1527.
Rao, S.M., and P. Shivananda. 2005. Role of osmotic suction in swelling of salt amended clays. *Canadian Geotechnical Journal* 42: 307–315.
Rao, S.M., and A. Sridharan. 1993. Environmental geotechnics—A review. *Indian Geotechnical Journal* 23: 235–252.
Rao, S.M., and T. Thyagaraj. 2007a. Role of direction of salt migration on the swelling behaviour of compacted clays. *Applied Clay Science* 38: 113–129.
Rao, S.M., and T. Thyagaraj. 2007b. Swell compression behavior of compacted clays under chemical gradients. *Canadian Geotechnical Journal* 44: 20–532.
Rao, S.M., T. Thyagaraj, and H.R. Thomas. 2006. Swelling of compacted clay under osmotic gradients. *Geotechnique* 56: 707–713.
Rao, S.M., T. Thyagaraj, and P. Raghuveer Rao. 2013. Crystalline and osmotic swelling of an expansive clay inundated with sodium chloride solutions. *Geotechnical and Geological Engineering* 31: 1399–1404.
Rao, S.M., G.B. Deepak, A. Anbazhagan, and P. Raghuveer Rao. 2016. Influence of physico-chemical components on the consolidation behavior of soft kaolinites. *Acta Geotechnica*. https://doi.org/10.1007/s11440-016-0478-0.
Saiyouri, N., D. Tessier, and P.Y. Hicher. 2004. Experimental study of swelling in unsaturated compacted clays. *Clay Minerals* 39: 469–479.
van Olphen, H. 1963. *An introduction to clay colloid chemistry*. New York: Wiley.
Villar, M.V. 2007. Water retention of two natural compacted bentonites. *Clays and Clay Minerals* 55: 311–322.
Yong, R.N. 1999. Soil suction and soil-water potentials in swelling clays in engineered clay barriers. *Engineering Geology* 54: 3–13.
Yong, R.N., and B.P. Warkentin. 1975. *Soil properties and behaviour*. New York: Elsevier.

Modern Geotechnical Practices

V. Raju and J. Daramalinggam

1 Introduction

The practise of geotechnical engineering has changed over the years. Certain practises which were once common are now deemed inefficient, unsafe, too polluting or too resource intensive. All aspects of geotechnical engineering have been impacted. Design methods, contracting frameworks, construction practices, equipment, construction materials, instrumentation and monitoring, data management, etc. have all changed.

This paper presents a small selection of modern geotechnical practices from around the world.

2 Features of Modern Geotechnical Design

Among other things modern geotechnical design is driven increasingly by the following:

- Advances in geotechnical design methods and computational power
- Updated codes of practise or standards
- Different contracting models (e.g. Design and Construct)
- Increased industry competition
- Advances in machinery
- Development of new techniques

V. Raju (✉)
Keller Group Plc, London, UK
e-mail: venu.raju@keller.co.uk

J. Daramalinggam
Keller AsiaPacific Ltd, Singapore City, Singapore

K. Ilamparuthi and R. G. Robinson (eds.), *Geotechnical Design and Practice*,
Developments in Geotechnical Engineering,
https://doi.org/10.1007/978-981-13-0505-4_4

Fig. 1 Photo of CPT Rigs at Changi Airport extension project, Singapore

2.1 Improved Soil Investigation Data

With the improvement of soil investigation technology, a variety of tools are available to the engineer to obtain design input. For example, while in the past Standard Penetration Test results may have been the only in situ test data available, we are seeing increasing use of Cone Penetration Tests (Fig. 1). Borehole sampling methods have also improved (e.g. piston samplers for soft clays, open drive samplers for stiff clays, Vibrocore samplers for sands) leading to better laboratory data. Geophysical testing (e.g. ground penetrating radar, seismic surveys, resistivity surveys) are slowly being adopted to supplement traditional boreholes and CPTs. Developers and Owners have also begun to specify more SI, thereby increasing the information available to designers and reducing uncertainty and reducing overall cost. Improved SI data is critical also to a designer's ability to effectively use modern design tools such as finite element methods.

2.2 Increased Use of Numerical Modelling Tools

Empirical, semi-empirical and analytical methods of design (for example Priebe's (1995) method for Vibro stone columns) remain relevant and in common use. Yee et al. (2015) document a large scale "zone test" in Malaysia on Vibro stone columns. Priebe's (1995) method was used to successfully predict the magnitude of settlements, and Han and Ye's (2001) method was used to successfully estimate the rate of consolidation settlements. The key was *locally calibrated parameters*. Indeed, many of these "older" design methods have been calibrated by a wealth of data, and have been refined over the years.

However, there are design problems where only a more sophisticated analysis yields a more accurate answer. For example, in a project in Malaysia, a 15 m high iron ore stockpile was seated on ground improved by Vibro stone columns, but the

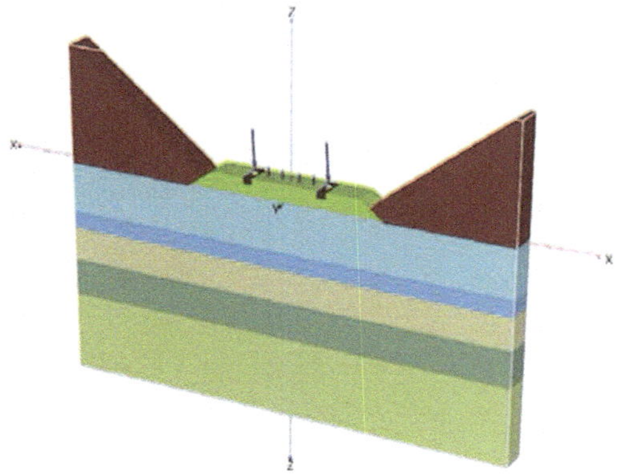

Fig. 2 Overview of a typical 3-D slice

stacker–reclaimer was founded on large bored piles. This created a complex geometry and loading sequence where only a 3-D finite element model could replicate the necessary components of the problem (Fig. 2). Engineers did not model individual stone column, as the focus was on the bored piles. Instead they used a composite material to model the stone column improved ground, but modelled the piles as solid elements. The more accurate 3-D analysis was used to determine the bending moments and deflections of the bored pile (Fig. 3). In contrast, a 2-D finite element model predicted bending moments that were 20–30% higher. However, as with all design methods, finite element models have to be calibrated and updated against actual data.

2.3 *The Use of the Observational Method*

Although the Observational Method (Peck 1969) is not new, its use is made more interesting and effective when coupled with modern instrumentation and design tools. An interesting use of updated parameters based on actual data was demonstrated by Low and Ng (2014) in Singapore. They report the case of the deletion of the sixth level of struts at the DTL Contract C922 overrun tunnel as part of Singapore MRT expansion. Due to delays and a tight project schedule, the main contractor decided to look at the possibility of omitting strut level S6 in a 23 m wide, 440 m long and 25 m deep cut-and-cover tunnel. When the excavation reached the third and fourth level of struts, the measured wall movements and strut forces were back-analysed to determined more realistic soil parameters for the various layers in the Old Alluvium formation. Compared to the original design, stiffness values had to be increased to match the measured wall movements. Importantly, the designers realised that an undrained Mohr Coulomb soil model (for the Old Alluvium layers) was the best match for the data. The updated undrained analysis was then used to predict the wall movements, wall bending moments and strut forces if the sixth level of struts were to be omitted. Deflections, etc. were

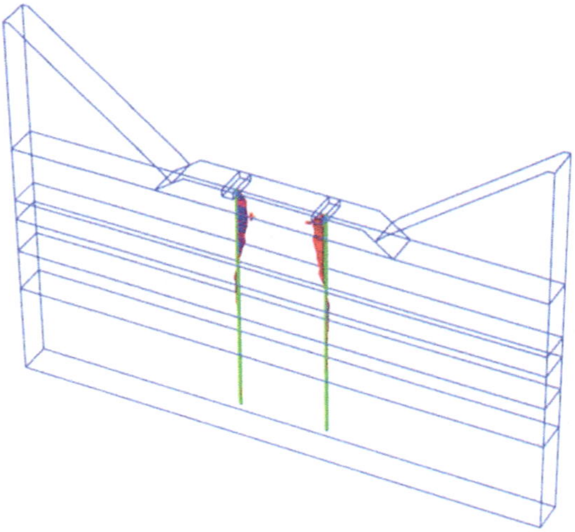

Fig. 3 Cut-away showing bending moments on piles with both sides loaded

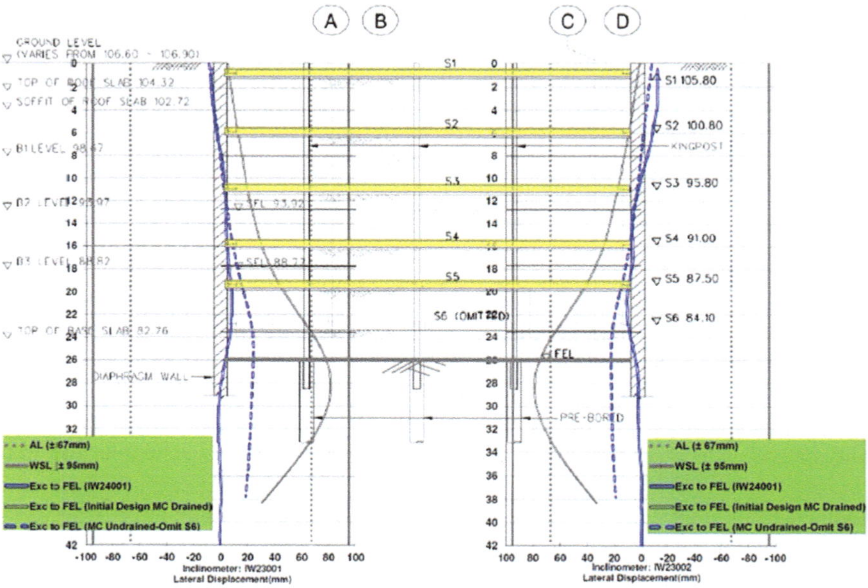

Fig. 4 Comparison of initial design (MC drained), revised design (MC undrained, without strut S6), measured deflections, alert level and work suspension level (Low and Ng 2014)

judged to be acceptable and construction proceeded without the sixth level of struts. Of course, as excavation proceeded, wall movements and strut forces were measured against the updated design (Fig. 4).

An important observation is that in this case, the use of an undrained analysis, even with more optimistic parameters, more closely approximated the measurements. This indicates the conservatism of using a drained analysis, which is common in Singapore for excavations in Old Alluvium.

2.4 *Acceptance of Alternative Solutions*

A leading shipyard, offering ship repair, shipbuilding, ship conversion, rig building and offshore engineering services developed a new shipyard located in Tuas View extension, Singapore. The development includes construction of dry docks, a hull block assembly workshop and other structures. The site investigation reports revealed that the sub-soil consisted of about 30 m of recently placed reclamation fill. This reclamation fill comprised mostly of sand fill with up to 12 m sandy clay. A marine clay of 9 m thick was also found in the south-eastern part of the area where the workshop would be constructed. The initial proposal was to use driven piles, but because of the potential negative skin friction over part of the site, the solution became very costly and time consuming. Therefore the owner was open to alternative solutions.

Keller proposed the Vibro compaction method for the sands and the installation of Prefabricated Vertical Drains (PVDs) with surcharge to treat the clays (Fig. 5).

PVDs were installed to a depth of 25–40 m with a 5.6 m high surcharge in the area where the sandy clay and marine clay layers were found. The construction of PVDs in an area over 30,400 m^2 was completed one and a half months. Vibro compaction works (Fig. 6) were carried out to improve the sandy layer that ranged from 15 to 28 m. An area of more than 126,400 m^2 was treated.

Surcharging was carried out to achieve 90% degree of consolidation and the settlement was monitored using the settlement plates, extensometers and magnetic spiders. Results were plotted using an Asaoka (1978) hyperbolic plot to check if the settlements were happening as planned. Figure 7 shows the plot of settlement versus time, both measured and predicted.

As the ground improvement was completed and the structure was built, monitoring continued to see if settlements were behaving as expected. Both settlement

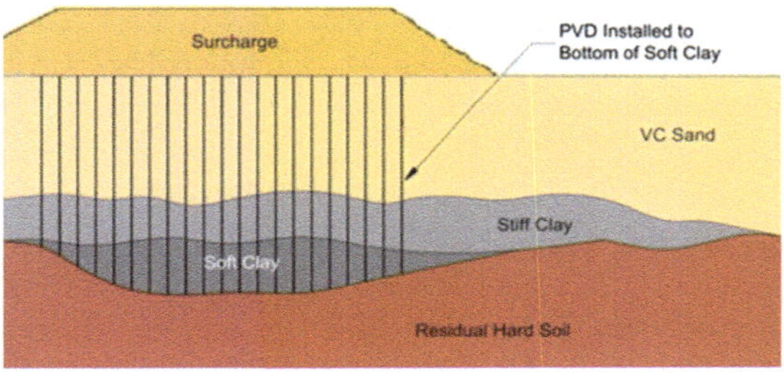

Fig. 5 Schematic of proposed ground improvement approach

Fig. 6 Vibro compaction works in progress, as surcharge is being placed

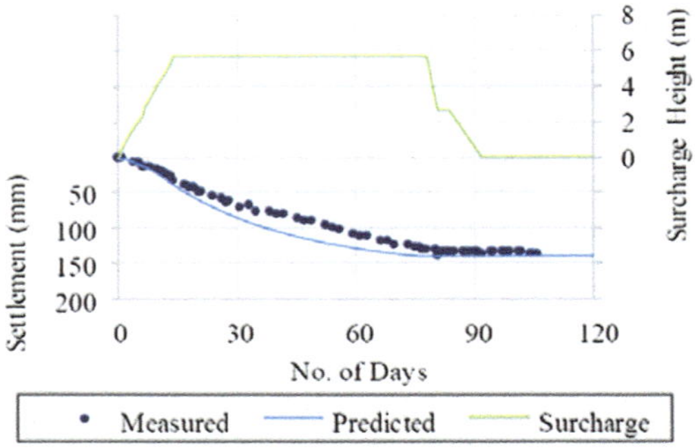

Fig. 7 Plot of predicted and measured settlements with time

magnitudes and settlements rates were monitored. The workshop is currently operational and settlements are under the limit of 50 mm. This project is described fully by He et al. (2016).

3 Equipment

3.1 Deeper, Faster, More Powerful, Automated

As infrastructure construction moves forward, for example huge airports which can handle more than 100 million passengers a year, the need to build on unfavourable soil conditions has increased. This in turn requires equipment which is able to penetrate to larger depths, work much faster to higher quality standards—which often requires a high degree of automation. An example of such equipment is the Keller Vibrocat TR07, for the installation of Vibro stone columns (see Fig. 8).

TR07 is latest in the series of Vibrocats designed and manufactured by KGS, the equipment wing of Keller. The TR07 is powered by 605 HP engine, able to generate 25 tonne pull-down force. Coupled with a mast height of 28.5 m and penetration depth of 23 m, TR07 is a truly modern rig. The increased pull-down

Fig. 8 Vibrocat TR07 working in Aachen, Germany (2013)

capacity and penetration depth enables TR07 to operate in tougher soil condition and reach greater depths than its predecessor. TR07 is designed to be safer and more user friendly, with an option for a fully automated column construction process. TR07 is equipped on-board with an air compressor, generator and hydraulic system, which makes it an all-in-one integrated machine. TR07 is also fitted with the latest M5 data acquisition systems which enable a user in any part of world to log into the system and analyse the data collected during installation of stone columns.

3.2 Smaller, More Compact, Versatile

Rigs are not just getting bigger. With the trend towards underpinning existing buildings, there is a need for smaller, more user-friendly equipment. Keller's KB1 (Series 2) rig is a compact rig, able to "fold" to just 1.85 m height and 0.76 m width-able to drive through a standard doorway. This allows it to access basements and other small spaces, where underpinning work is required (Fig. 9).

Fig. 9 KB1 rig, accessing a hard-to-reach area

3.3 Data Acquisition and Management

Modern equipment now comes with a variety of sensors that provides real-time feedback to the rig operator for improved productivity, safety and quality of work. All major rig manufacturers now equip their piling or small drilling rigs with computer systems that measure and display hydraulic pressures, rotation speeds, mast inclination and other key parameters.

In the USA, Hayward Baker, a Keller company, has gone a step further, outfitting its wet Deep Soil Mixing (DSM) rigs with an in-house data acquisition system that complements the on-board systems supplied by the manufacturer (Fig. 10).

To create a continuous DSM "sealing" slab (Fig. 11) a very basic concern is the accurate positioning the mixing tool at the centre of the column to be formed. Traditionally this has been done using survey pegs. However, in a DSM site, with closely spaced columns and slurry overflows, pegs can be easily disturbed and lost. Hayward Baker uses rig-mounted precision GPS to locate the centre of the columns to commence column construction. This eliminates the need for survey pegs. In addition, as the column is being constructed, key column parameters (binder flow rate, binder content, blade rotation number) are read, displayed to the operator and

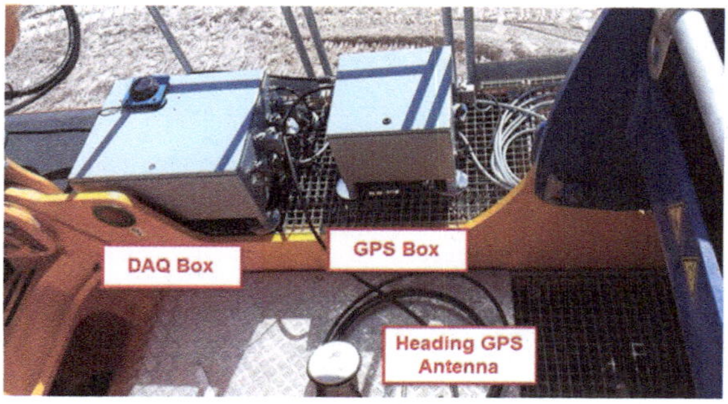

Fig. 10 Hayward Baker's DAQ system, being setup on a DSM rig

Fig. 11 Hayward Baker's DAQ system was employed on a job in Miami, USA, for the construction of a sealing slab close to the beach, with stringent positional tolerances

captured in digital form (Fig. 12). This reduces human error and greatly reduces the administrative work for the site engineer, who is able to use the automatic outputs for quality control purposes.

4 Improved Safety and Environmental Standards

4.1 General Construction Safety

Since 1 August 2016, the "Design For Safety" legislation came into force in Singapore, covering projects over S$ 10 million (approximately INR 500 million).

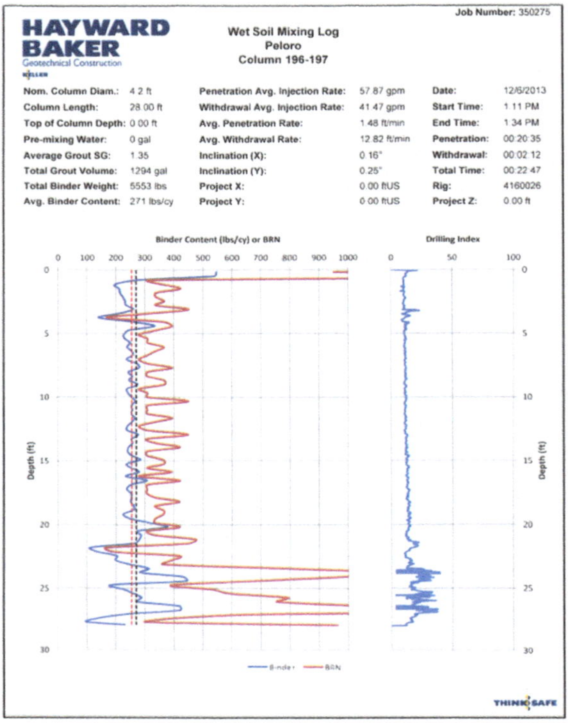

Fig. 12 Sample construction printout from DSM rig

The legislation requires Developers to work with Designers to ensure that foreseeable design risks are eliminated or reduced to "As Low As Reasonably Practicable", and that residual risks are managed. Contractors are also responsible to keep the Developer or Main Contractor informed of any risks they foresee. A key person in this process is the DfS (Design for Safety) Professional, whose responsibility it is to convene DfS meetings on behalf of the Developer and maintain an updated copy of the DfS Register. This register will be handed over to the Owner for future reference and use.

This legislation is important, because many risks faced by workers on site are far more efficiently managed if anticipated at design stage. A key principle underlying this legislation is that collective protection measures (e.g. covered walkway) are preferable to individual protective measures (e.g. safety helmet) Indeed, some risks can only be mitigated at design stage. For example, the sequence of strut or anchor installation in a deep excavation is a geotechnical and structural consideration that has a significant impact on the safety of the work on site, but can only be addressed at design stage.

There has also been increased adoption of modern risk-based safety management systems, for example OHSAS 18001: 2007, which will soon be updated. While many of these legal requirements and practices are first adopted by larger main contractors, geotechnical engineering contractors are following suit.

4.2 Safety in Geotechnical Engineering

4.2.1 Safer Equipment

In the European Union, legislation is in place to require rig manufacturers and contractors to instal and use machine guarding that protects workers from accidental contact with rotating parts-which pose a significant risk of fatalities or major injuries.

Modern foundation equipment and drill rigs now come equipped with Emergency Stop switches, cameras in critical positions, load limiters, etc. and are generally designed to be easier to maintain and repair. For example, the Liebherr LB 36 piling rig is designed to be safely assembled without the use of an auxiliary crane, and has a cabin that is designed to give the rig operator maximum visibility of his surroundings, several cameras covering the operators "blind spots", and built in walkways and railings for safe access of the upper-carriage for maintenance and repair. The design also minimises noise in the cabin (less than 78.3 dBA) and minimises vibrations to the hand and body or the rig operator.

As equipment fleets are renewed, and modern equipment is brought into replace old equipment, contractors should factor in safety features in their decision-making process.

4.2.2 Safer Construction Methodologies

With geotechnical contractors adopting risk-based management systems, construction methods are often modified to reduce the risk to people and equipment. A contractor will conduct a risk assessment of all activities (consisting of hazard identification, risk assessment and risk mitigation) before starting work. For example, because of the risk of excavation collapse during pile hacking and pile cap construction, excavation pits are sloped at an appropriate angle or sheet piles are used, checked by a geotechnical engineer and not just left to the site team to decide. In the past, these decisions may have been left to the site foreman.

4.3 Consideration of Environmental Impact

Increasingly, owners are looking not only for the lowest price, but also for a geotechnical solution that minimises impact on the environment. Some contractors are themselves looking for solutions that do the same, driven partly by investor and

Fig. 13 Overview of Srejach Tunnel site

public sentiment. Minimisation of environmental impact can take many forms. These include (a) more efficient use of materials, often resulting in cost and time savings also, (b) use of recycled materials in construction, (c) reduction of polluting processes, and (d) reduction of noise, dust and vibrations.

The Austrian railway authorities are building a new 130 km line from Klagenfurt to Graz that will cut travel time from the existing 2 h to a mere 45 min. Approximately 50 km of this new line is underground. Keller was involved in performing jet grouting work at two tunnel sections.

At the Srejach Tunnel (Fig. 13), a 600 m long cut-and-cover tunnel, jet grouting was performed for a stiffening slab and to seal gaps between bored piles. In total nearly 7,000 columns were constructed. Jet grouting works were completed in August 2015. At the Untersammelsdorf Tunnel, a 600 m long mined tunnel, jet grouting commenced in September 2016 and is being performed to a maximum depth of 28 m, sometimes drilling through 25 m of overburden.

At this tunnel a total of about 7,500 columns will be constructed for stiffening slabs and for sealing between bored piles.

For this job, large jet grouting equipment is being mobilised. Two Techniwell TW700 pumps are being used for the work, in addition to two Keller "AKM" mixers, three 60 tonne cement silos, water tanks and of course jet grouting rigs. A key feature of this job is the use of a MAT B170 desander and Technioda DC1300 filterpress, for sludge management (Fig. 14). As jet grouting uses the erosive force of the fluid to cut and mix the soil, large quantities of sludge are produced. Depending on the soil type, anywhere between 1.5 and 3 times the volume of the jet grout element is created as sludge, which then needs to be disposed. Huge quantities of water are involved. Historically, this sludge was simply left to cure (because of the cement) and disposed by truck to a landfill. Keller used the desander and the filterpress to extract the water from the sludge for reuse, and simultaneously reduce the volume of sludge to be disposed.

The environmental benefits of this approach are clear. While some diesel is used to run the desander and filterpress, the amount of water being recycled is significant. In addition, there is a dramatic reduction in the volume of material needing disposal

Fig. 14 Overview of Desander and Filterpress setup

by three times, leading to reduce truck journeys and also reduction in landfill usage. At the Srejach Tunnel, 33,500 m^3 of dewatered sludge was produced in the operations, down from over 90,000 m^3 if the conventional methods were used.

5 Some Future Trends

Sondermann and Kummerer (2016) highlight the following global trends that will have an impact on the construction sector.

- An increased proportion of construction in developing countries as their economies rapid outpace developed countries
- Continued growth of mega-cities, particularly in Asia
- Increased demands on energy production
- An increased trend towards private funding of construction work, and public funding diminishes in relative terms

Sondermann and Kummerer (2016) further suggests that

- Construction projects will grow in complexity and scale
- Competition will become more and more global
- There will be an increase of automation in both construction and monitoring
- There will be increased demands for documentation
- Clients will demand more environmentally friendly processes

It is therefore prudent and indeed necessary for players in the geotechnical industry to learn from modern practices across the world, and adapt it to local conditions.

References

Asaoka, A. 1978. Observational procedure of settlement prediction. *Soils and Foundations* 18 (4): 87–101.

BS OHSAS 18001: 2007. *Occupational health and safety management-requirements*. BSI.

Han, J., and S.L. Ye. 2001. Simplified method for consolidation rate of stone column reinforced foundations. *Journal of Geotechnical and Geoenvironmental Engineering* 127 (7): 597–603.

He, Z.W., E. Lin, K.W. Leong, and V.R. Raju. 2016. Design and performance of ground improvement for a ship hull workshop in Singapore. In *Proceedings of the 19th Southeast Asian geotechnical conference and 2nd AGSSEA conference*, ed. S.H. Chan, T.A. Ooi, W.H. Ting, S.F. Chan, and D. Ong, 469–473.

Low, S.Y.H., D.C.C. Ng. 2014. Omission of strut by observation approach to speed up construction of DTL Contract C922 Overrun Tunnel. In *Proceedings of underground Singapore 2014*, 224–247.

Peck, R.B. 1969. Advantages and limitations of the observational method in applied soil mechanics. *Geotechnique* 19 (2): 171–187.

Priebe, H.J. 1995. The design of vibro replacement. *Ground Engineering* 28 (10): 31–37.

Sondermann, W., and C. Kummerer. 2016. Future challenges for geotechnical engineers (a more contractor oriented perspective). In *Challenges and innovations in geotechnics*, ed. A. Zhussupbekov, 3–8. London: Taylor & Francis.

Yee, Y.W., P.V.S.R. Prasad, L.H. Ooi, and J. Daramalinggam. 2015. *Instrumented low embankment on stone columns for the Ipoh-Padang Besar double tracking project*, 23–25, Jurutera, January 2015. The Institution of Engineers Malaysia.

A Design Framework for Spatial Variability in Cement-Treated Soft Clay in Deep Excavations and Underground Constructions

Fook-Hou Lee, Yong Liu, Akanksha Tyagi, Kai-Qi Tan, Yutao Pan and Jiahui Ho

1 Introduction

In tunnelling and underground construction in areas underlain by soft soils, ground improvement works is often undertaken. A widely used approach for ground improvement is cement admixture, wherein cement powder or slurry is mixed into the soft soil, either using deep cement mixing or jet grouting, to change the soft ground into a hardened mass. Significant heterogeneity has been observed in such cement-treated soil. For instance, Chew et al. (1997) presented jet grouting core sample data which showed unconfined compressive strength varying from less than 1 MPa to about 4 MPa. Similarly, Chen et al.'s (2011) field data also indicate unconfined compressive strength of core samples ranging from approximately 0.7–5 MPa for the deep mixing work at the Marina Bay Financial Centre (MBFC) project in Singapore. For dry lime mixing, Larsson et al. (2005) also noted similarly significant spatial variation. This range of spatial variation in strength is significantly larger than that encountered in many naturally occurring soils. Chen et al.'s (2016) data from the Marina Bay Financial Centre and Marina One as well as data from other cement treatment projects conducted in Singapore indicate a coefficient of variation (COV) of the unconfined compressive strength typically ranging from 0.37 to 0.47. Chen et al. (2016) also noted that this range of COV can be derived from a corresponding COV for cement slurry concentration of roughly 0.2, which implies reasonably good mixing quality. Given the constraints of time and site conditions, it may be difficult to obtain a COV for slurry concentration which is less than 0.2. This would also mean that, in practice, it is difficult to reduce the COV in strength to much less than 0.4.

F.-H. Lee (✉) · Y. Liu · A. Tyagi · K.-Q. Tan · Y. Pan · J. Ho
Department of Civil & Environmental Engineering, National University of Singapore, Singapore City, Singapore
e-mail: leefookhou@nus.edu.sg

K. Ilamparuthi and R. G. Robinson (eds.), *Geotechnical Design and Practice*, Developments in Geotechnical Engineering,
https://doi.org/10.1007/978-981-13-0505-4_5

While significant spatial variability in cement-treated ground has been well-reported, there is still not well-established design method which has be shown to be able to account for such variability. In Eurocode 7 (e.g. Simpson and Driscoll 1998; Schneider 1999; Orr 2000; Eurocode 7 CEN 2004; Hicks 2013), the design strength is evaluated by applying partial factors to the mean strength. These factors are intended, in part, to account for spatial variability in strength (e.g. Simpson and Driscoll 1998). However, it is uncertain if these factors are applicable to cement-admixed soil and a more common approach is to adopt a design strength that is several times lower than the laboratory-measured strength using the same mix proportion (e.g. Nakagawa et al. 1996). In Singapore, a highly prescriptive variant to this approach is often used, wherein the design unconfined compressive strength of cement-treated clay is limited to about 700 kPa. This is much lower than the mean strength, which is typically in the range of 1.5–2 MPa. The low prescribed strength is likely to be a safety measure taken to account for the uncertainties and unknowns related to cement-treated clay. However, using such a low strength has unavoidable cost implications. Where the volume of ground treatment is substantial, it may be necessary to seek a more optimal balance between safety and build ability.

The issue of spatial variability and its effect on the design of ground improvement work has been receiving increasing attention in recent years. In particular, researchers at the National University of Singapore have been studying the effect of spatial variability of cement-treated soil on its mass performance in several different scenarios. This paper summarizes some of the research findings for underground cement-treated soil layer for deep excavations and improved soil rings around tunnels as well as design framework which these findings point to.

2 Underground Cement-Treated Soil Layer

A common concern in deep excavation projects in densely built-up environments overlying soft clayey soil is excessive wall deflection and ground movement, in particular settlement, around the excavation. Above the excavation level, struts can be installed to limit wall deflection. However, in soft clay soil conditions, the maximum wall deflection usually occurs below final excavation level, where struts cannot be installed. In such situations, cement admixture by deep mixing or jet grouting is often used to create cement-treated soil layers to act as underground props and thereby control wall deflection (e.g. Nakagawa et al. 1996; McGinn 2003). Cement-admixed soil layers are often assumed to be an isotropic and homogeneous Mohr–Coulomb material in design (e.g. O'Rourke and McGinn 2006). In reality, the ground improved by jet grouting and deep mixing has a strongly columnar structure (e.g. Nakagawa et al. 1996), and possesses significant random spatial variation with a coefficient of variation (COV) around 0.4 (e.g. Larsson et al. 2005; Chen et al. 2011).

Liu et al. (2015) studied the effect of spatial variation in strength of cement-treated soil slab taking into account three types of variability, as follows:

(a) Deterministic trend of decrease in strength with radial distance from the column centre within each cement-treated soil column.
(b) Stochastic variation in strength within each column.
(c) Random errors in column positioning, as summarized in Table 1. Such positioning errors may result from eccentricity of the mixing equipment or off-verticality, as illustrated in Fig. 1. If the treated soil stratum lies deep underground, a small tilt can result in large positioning errors at depth. Positioning error was considered by prescribing a uniform random error in the column centre locations, subjected to a maximum deviation.

Liu et al. (2015) used a strain-softening Mohr–Coulomb model in which the strength parameter was represented by the undrained shear strength and the angle of friction was assumed to be zero. Strain-softening was modelled by decreasing the undrained shear strength as a function of the shear strain. The elastic modulus was assumed to be proportional to the peak undrained shear strength. Since the latter varied randomly with location, so did the modulus. Figure 2 shows a typical mesh from a random realization. The darker areas represent the stronger and stiffer areas. White areas represent untreated clay zones. In Liu et al.'s (2015) problem, the cement-treated slab was subjected to lateral compression at its two ends to simulate the forces generated by the inward movement of the retaining walls. The sides of the slab were assumed to be either completely free to move laterally or completely fixed.

Table 1 Summary of maximum allowable deviation from design positions (after Liu et al. 2015)

Mixing method	Max. allowable deviation from design positions		References
	Start position	Verticality (compared by drilling depth)	
Jet grouting	50 mm	2% for depths up to 20 m, different tolerances for greater depths	BS EN (2001)
	76.2 mm	1%	Geo-Institute, ASCE (2009)
	–	0.5% for depths up to 15 m, 1% for depths up to 30 m	Stoel (2001)
	50 mm	0.5%	Morey and Campo (1999)
	50 mm	0.5%	Passlick and Doerendahl (2006)
Deep mixing	–	1%	Rutherford et al. (2005)
	50.8 mm	2%	Puppala et al. (2008)
	50 mm	1%	Bahner and Naguib (2000)
	–	1%	Ryan and Jasperse (1989)
	–	1–2%	Larsson (2005)

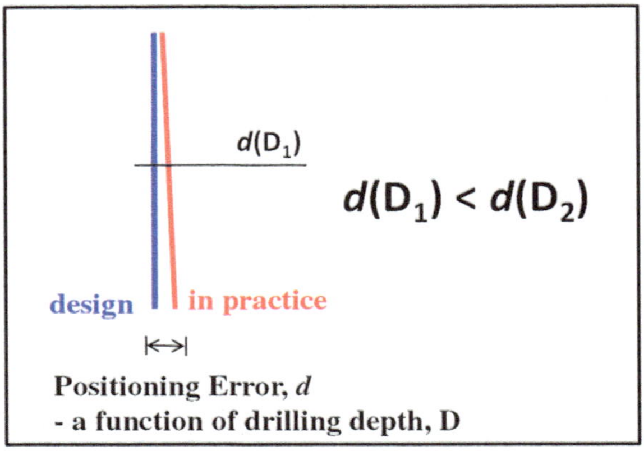

Fig. 1 Increase in positioning error with depth

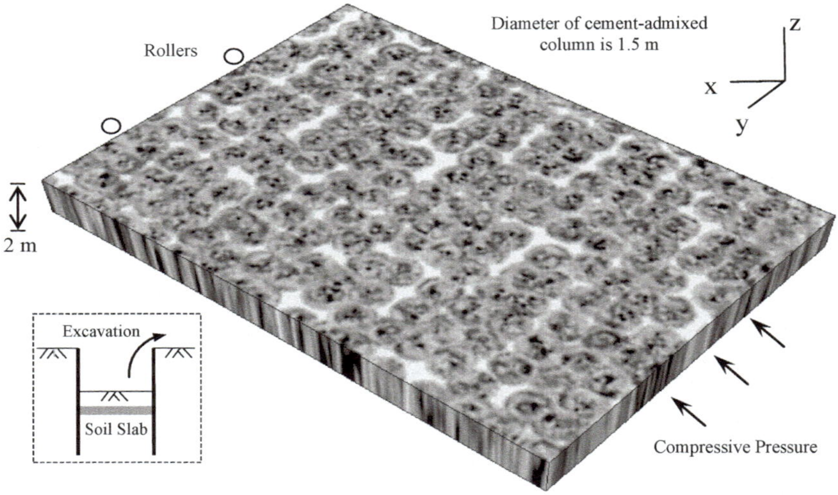

Fig. 2 A typical random realization of the cement-treated soil layer (after Liu et al. 2015)

In a random finite element analysis, an analysis is repeated many times, each termed herein as a realization. Since the soil properties at each point within the domain are randomly assigned following a set of prescribed statistical parameters, no two realizations are exactly alike. The results of all these realizations give a statistical spread of the outcome. Figure 3 shows a typical set of results from an analysis comprising 100 realizations. In this figure, the calculated stress is the average stress across the ends of the slab while the mass strain is the average strain,

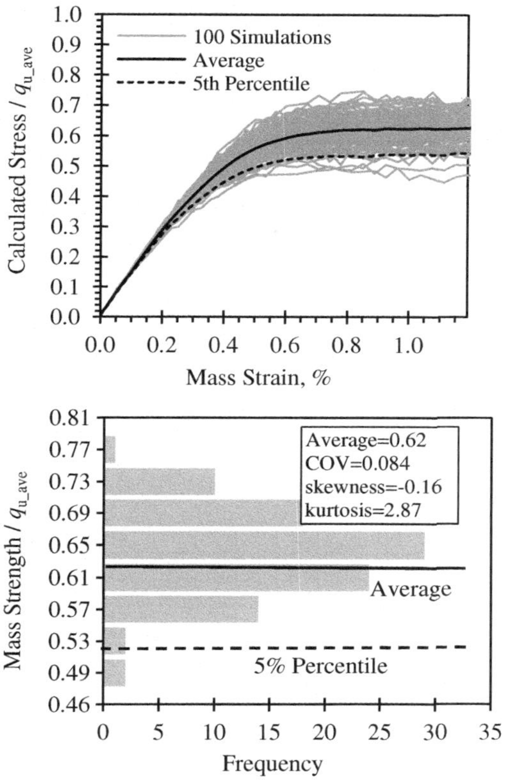

Fig. 3 (top) Typical mass stress-strain curves from 100 realizations; (bottom) histogram of failure stress (after Liu et al. 2015)

evaluated as the compression of the slab divided by its original length. As can be seen, the initial portion of each stress-strain curve is linear, thus allowing an equivalent 'mass modulus' to be evaluated. Furthermore, the end portion of each stress–strain curve is flat, thus allowing a 'mass strength' to be evaluated. As Fig. 3b shows, the distribution of the mass strength from the 100 realizations is reasonably close to a normal distribution. Based on this, Liu et al. (2015) proposed that a rational design strength Q_{d} that represents the performance of the entire system can be evaluated via the relationship

$$Q_{\mathrm{d}} = q_{u\_ave} - \alpha_Q \cdot \sigma_q \qquad (1)$$

in which q_{u_ave} is the volume-averaged strength, σ_q is the standard deviation of the strength variation in the treated soil and α_Q is a reduction coefficient which can estimated from the statistical properties of the mass strength distribution, coefficient of variation of the spot strength and the probability of failure. Details of the procedure for evaluating the reduction coefficient α_Q are given in Liu et al. (2015). For an unconfined compressive strength COV of 0.4, the reduction factor α_Q is approximately 0.6.

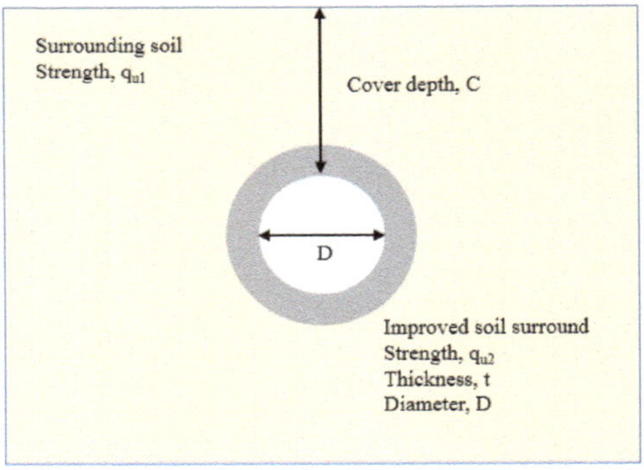

Fig. 4 Schematic of a tunnel with a cement-treated soil ring around it

3 Improved Soil Rings Around Tunnels

In tunnelling scenarios, where the soil is not fully self-supporting, cement-treatment is commonly used to create a zone of hardened soil around the tunnel periphery, hereafter termed improved soil surround, to prevent soil collapse (e.g. O' Rourke and McGinn 2006), as illustrated schematically in Fig. 4. Such improved soil concepts include the jet-grouted 'safe haven' box used in the SR99 four-lane bored tunnel, Washington (WSDOT 2016), the jet-grouted dome atop the Kanadská tunnel, Prague (Zakladani 2012) and Aeshertunnel, Switzerland (Coulter and Martin 2004). Similar improved soil surrounds are also being used in contract T227 of the Thomson Line, currently under construction in Singapore. The presence of improved soil surround is likely to alter the failure mechanisms and stability characteristics of the tunnel significantly from that before improvement. Although field observations on tunnels in improved soil surrounds have been reported (e.g. Canetta et al. 1996; Pellegrino and Adams 1996), systematic studies on the failure behaviour and stability characterization of tunnels with improved soil covers or surrounds have generally been scanty. Indeed, much remains unclear about the failure modes of tunnels with such improved soil surrounds.

The problem of an improved soil ring around a circular tunnel is much more complex than that of a cement-treated soil slab because the stress states induced by the loading are much more non-uniform. Centrifuge model tests and deterministic finite element analyses indicate that weaker and thinner cement-treated soil rings tend to fail by shearing of the crown region of the soil ring, accompanied by plastic collapse of the soft soil above it. Thicker and stronger cement-treated soil rings tend to fail by tension cracking instead, Fig. 5. In such failures, the surrounding soil movement is typically much smaller and surface settlements cannot be used as an indicator of cracking. Instead, stress transducers were embedded at the crown of the soil ring extrados to monitor the contact stress between the cement-treated soil ring and the soft ring (Zulkefli et al. 2015). Cracking of the soil ring gave rise to a small settlement of the crown of the soil ring which is registered as a decrease in the contact stress.

Fig. 5 Cracking of a cement-treated soil ring in centrifuge model test. Unconfined compressive strength 1 MPa

Random finite element analysis was conducted on the cement-treated soil ring using an effective stress Mohr–Coulomb model. The friction angle was based on an average value from triaxial tests while the effective cohesion was estimated from the unconfined compressive strength. Xiao (2009) observed water being expelled from the sides of unconfined compressive test specimens and postulated that, for very stiff material such as cement-treated soil with high cement, the unconfined compressive test approximates a drained, rather than undrained test. This allows the effective stress states of an unconfined compressive test specimen to be defined and the effective cohesion to be estimated. Since the friction angle is constant, the effective cohesion can be shown to be linearly related to the unconfined compressive strength.

The software used for the finite element analyses is GeoFEA9.0 (GeoSoft 2016). The random material properties were incorporated into GeoFEA as a user-defined material subroutine. Tension cracking was not explicitly modelled. Much remains unknown about propagation of tension cracks in cement-treated soil. For a spatially and randomly variable materials, the challenges of modelling crack propagation are even more formidable. Instead, a phenomenological approach is adopted in this study. Centrifuge observations indicate that cracking typically occurs at the crown, invert and shoulders of the tunnel. In the random finite element analysis, the tensile stresses were monitored for the cement-treated soil ring having spatially variable strength and modulus. Pan et al. (2016) reported that the tensile strength of cement-treated marine clay is typically about 13% of the unconfined compressive strength. This was then used as a criterion to define cracking failure. The excavation of the tunnel was carried out till a specified out-of-balance (OOB) force exceedance (10 or 20%). This failure criteria was established from deterministic results in which shear/tension failure, tunnel volume loss and tunnel distortion were observed once the out-of-balance force exceeds 10–20%. Figure 6 shows the critical tunnel support pressure from a random analysis. The critical tunnel support pressure is the tunnel support pressure at which failure occurs and it is expressed as a percentage of initial tunnel support pressure. As can be seen, the distribution of critical tunnel

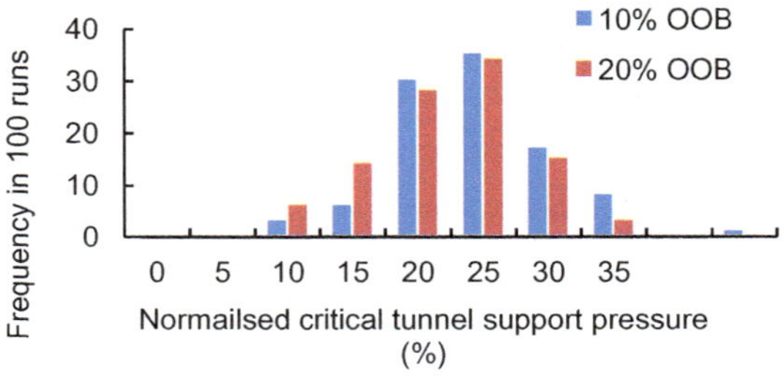

Fig. 6 Critical tunnel support pressure from random analyses

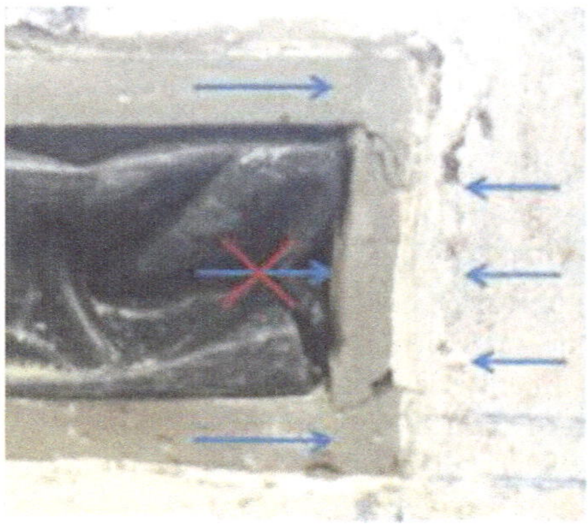

Fig. 7 Failure of cement-treated soil heading in centrifuge model test

support pressure is also approximately normal. The analyses are still on-going. Results to-date suggest that, for a probability of failure of 0.05 (corresponding to the 5th percentile of the distribution) and unconfined compressive strength COV of 0.4 and ignoring positioning error, the value of the reduction coefficient α_Q is about 1.35. This is significantly higher than that obtained by Liu et al. (2015) for cement-treated soil slab. The difference may be attributed to the fact that the stresses due to the external loading are much more non-uniform for a cement-treated soil ring. For lower probability of failure, the reduction coefficient will increase. For higher unconfined compressive strength COV, the reduction coefficient α_Q may decrease slightly, because it is more than offset by the increase in standard deviation, Eq. (1). However, it may be prudent to ignore the decrease and assume the same value for reduction coefficient.

Similar analyses were also conducted for cement-treated tunnel headings, Fig. 7. Centrifuge model tests and deterministic analyses indicate such tunnel headings fail predominantly by diagonal tension cracking along the circumference. Random finite element analyses were also conducted using a similar approach as that described above for cement-treated soil ring. Failure was considered to have occurred when the minor principal effective stress in the zone around the entire periphery of the heading falls below the negative of the tensile strength of the soil.

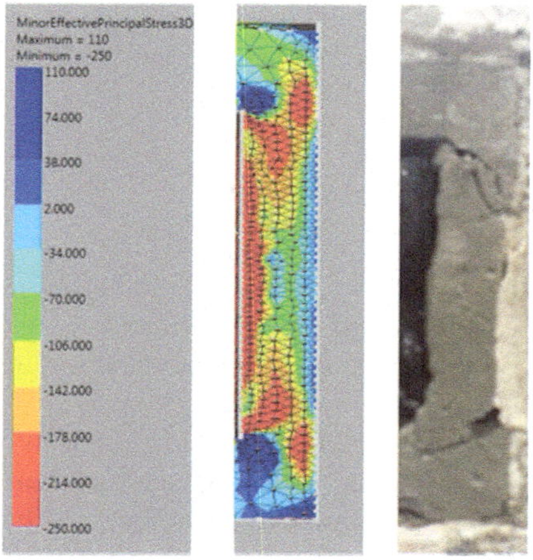

Fig. 8 Minor principal effective stress contours and crack locations at heading

This is also supported by deterministic finite element results, Fig. 8. Results to-date suggest that, for a probability of failure of 0.05 (corresponding to the 5th percentile of the distribution) and unconfined compressive strength COV of 0.4 and ignoring positioning error, the value of the reduction coefficient α_Q is about 1.5. This is slightly higher than the value of 1.3 for cement-treated soil ring discussed earlier.

4 Effect of Constitutive Model

The analyses above were all conducted using Mohr–Coulomb model. Xiao et al. (2014) showed that cement-treated clay undergoes significant loss of structure after yielding. Under undrained triaxial conditions, strain-softening takes place shortly after initiation of yielding. In drained triaxial conditions, strain hardening tends to occur, accompanied by large volumetric compression. These behavioural features are not reflected by the Mohr–Coulomb model. At the same time, however, the Mohr–Coulomb is the most widely model in practical geotechnical engineering, including cement-treated soil. Advanced models for structured soils may not come into wide usage in the near future. For this reason, there is a need to assess the errors involved in the use of Mohr–Coulomb model.

To do this, comparison were also made between the Mohr–Coulomb model and Xiao et al.'s (2016) cohesive Cam Clay model. The results of the comparison showed that, for deterministic problems or problems involving homogeneous domains, the failure load from the Mohr–Coulomb model is usually about 10% higher than that obtained using the cohesive Cam Clay model. However, the cohesive Cam Clay model often gives a wider spread of failure loads in random analyses. Comparison of the 5th percentile results indicates that the Mohr–Coulomb model may underestimate the reduction factor by up to 25%, compared to the cohesive Cam Clay model. As such, more studies are needed to shed further light on the interaction between model and spatial variability.

5 Conclusion

The foregoing discussion indicates that a feasible of analysing spatial variable cement-treated soil constructions in practical design is to lower the mean strength by a multiple of the standard deviation of the cored strength; this multiple being termed above as a reduction factor. The results suggest that the value of the reduction factor depends upon the stresses arising from external loading. For highly uniform stress situations such as that in a cement-treated soil slab, the reduction factor is relatively low. On the other hand, for non-uniform stress situations such as those in cement-treated soil rings and heading, the reduction factor is substantially higher. This implies that the non-uniform stress distribution can aggravate the effect of material spatial variability, and vice versa. An issue which is still receiving on-going attention is the interaction between material model and spatial variability. Results to-date indicate that the Mohr–Coulomb model tends to err on the unsafe side. If this is so, then Mohr–Coulomb parameters may need to be factored down accordingly.

References

Canetta, G., B. Cavagna, and R. Nova. 1996. Experimental and numerical tests on the excavation of railway tunnel in grouted soil in Milan. In *Geotechnical Aspects of Underground Construction in Soft ground*, ed. Taylor Mair, 479–484. Rotterdam: Balkema.

CEN. 2004. *EN 1997-1: 'Eurocode 7—Geotechnical design—Part 1: General rules'*. Brussels: European Committee for Standardization.

Chen, J., F.H. Lee, and C.C. Ng. 2011. Statistical analysis for strength variation of deep mixing columns in Singapore. *GeoFrontier 2011: Advances in Geotechnical Engineering* 576–584.

Chen, E.J., Y. Liu, and F.H. Lee. 2016. A statistical model for the unconfined compressive strength of deep-mixed columns. *Geotechnique*, 66 (5): 351–365.

Chew, S.H., F.H. Lee, Y. Lee, and I. Yogarajah. 1997. Jet grouting in Singapore Marine Clay'. In *413 Proceedings of 3rd Young Geotechnical Engineers' Conference*, ed. T.S. Tan, S.H. Chew, K.K. Phoon, and T.G. Ng, 231–238. Singapore.

Coulter, S., and C.D. Martin. 2004. Ground deformations above a large shallow tunnel excavated using jet grouting'. In *Proceedings of ISRM Regional symposium, EUROCK 2004 and 53rd Geomechanics Colloquy*, ed. W. Schubert, 155–160. Essen: VGE.

GeoSoft. 2016. http://www.geosoft.sg/products/geofea.php.

Hicks, M.A. 2013. An explanation of characteristic values of soil properties in Eurocode 7. In *Modern geotechnical design codes of practice*, ed. P. Arnold, G.A. Fenton, M.A. Hicks, and T. Schweckendiek, Simpson, 36–45. IOS Press.

Larsson, S., H. Stille, and L. Olsson. 2005. On horizontal variability in lime-cement columns in deep mixing. *Géotechnique* 55 (1): 33–44.

Liu, Y., F.H. Lee, S.T. Quek, E.J. Chen, and J.T. Yi. 2015. Effect of spatial variation of strength and modulus on the lateral compression response of cement-admixed clay slab. *Géotechnique* 65 (10): 851–865.

McGinn, A. J. 2003. *Performance of deep excavations in Boston marine clay stabilized by deep mixing methods*. Ph.D. thesis, Cornell University.

Nakagawa, S., I. Kamegaya, K. Kureha, and T. Yoshida. 1996. Case history and behavioral analyses of braced large scale open exaction in very soft reclaimed land in coastal area. In *Proceedings of the International Symposium on Geotechnical Aspects of Underground Construction in Soft Ground*, 179–184. London.

O'Rourke, T.D., and A.J. McGinn. 2006. Lessons learned for ground movements and soil stabilization from the Boston Central Artery. *Journal of Geotechnical and Geoenvironmental Engineering, ASCE* 132 (8): 966–989.

Orr, T.L.L. 2000. Selection of characteristic values and partial factors in geotechnical designs to Eurocode 7. *Computers and Geotechnics* 26: 263–279.

Pellegrino, G., and D.N. Adams. 1996. The use of jet grouting to improve soft clays for open face tunnelling. In *Geotechnical Aspects of Underground Construction in Soft ground*, ed. Taylor Mair, 423–428. Rotterdam: Balkema.

Schneider, H. R. 1999. Definition and determination of characteristic soil properties. In *Proceedings XII International Conference on Soil Mechanics and Geotechnical Engineering*, vol 4, 2271–2274. Hamburg.

Simpson, B., and R. Driscoll. 1998. *Eurocode 7: A Commentary*. Watford, UK: Construction Research Communications.

WSDOT, Washington State Department of Transportation. 2016. Retrieved on February 11, 2016, from http://www.wsdot.wa.gov/Projects/Viaduct/About/Tunneling.

Xiao, H., F.H. Lee, and K.G. Chin. 2014. Yielding of cement-treated marine clay. *Soils and Foundations* 54 (3): 488–501.

Xiao, H., F.H. Lee, and Y. Liu. 2016. Bounding surface cam-clay model with cohesion for cement-admixed clay. *International Journal of Geomechanics, ASCE* 04016026.

Zakladani. 2012. Retrieved on February 11, 2016 from http://www.zakladani.cz/en/underground-line-v-a-track-section-dejvicka-bor-islavka-sod-02-securing-the-access-tunnel-kanadska-so-02-29-02.

Zulkefli F., E. Tan, F.H. Lee, and S.H. Goh. 2015. 1-g model tests of tunnels with a surrounding cement-treated soil ring. In *Proceedings of 15th Asian Regional Conference on Soil Mechanics and Geotechnical Engineering* (Paper SIN-11).

Understanding Performance and Developing Global Geotechnical Standards

Kancheepuram N. Gunalan

1 Introduction

Major infrastructure project investments are primarily focused on mitigating financial risk in meeting long-term performance criteria/requirements of specific elements of work. Long-term performance of infrastructure facilities rely heavily on a very diligent and a pragmatic design by the geotechnical engineer who not only understands how to design the various elements meeting the prescribed criteria but also has the experience to predict performance over the life of the contract so as to minimize long-term maintenance costs. We have all heard various geo legends say "a successful practitioner is one who exercises good judgement in addition to having a sound understanding of fundamental behavior of soils in a given environment".

This presentation provides a very brief overview on understanding and evaluating geotechnical performance criteria/requirements to develop global performance standards.

2 Background

New infrastructure to meet a growing demand or to reconstruct an aging infrastructure, all need financial resources to plan, design, construct, operate and maintain. Agencies around the globe are all facing revenue challenges to not only build new and improved infrastructure but to maintain existing ones in good state of repair. The challenge can be attributed to unwillingness by the relevant legislative body to raise taxes; escalation cost of money due to delayed procurement or exe-

K. N. Gunalan (✉)
AECOM, 756 East Winchester Street, Suite 400, Salt Lake City, UT 84107, USA
e-mail: k.n.gunalan@aecom.com

K. Ilamparuthi and R. G. Robinson (eds.), *Geotechnical Design and Practice*, Developments in Geotechnical Engineering,
https://doi.org/10.1007/978-981-13-0505-4_6

cution; impact due to increased material or labor costs, etc. Agencies are not only looking for creative ways to finance these projects but also to plan, design, construct, operate, and maintain them.

Typical process up until recently has been for the agency to plan, design, bid, and construct the projects also known as design-bid-build (D-B-B) delivery method. In a D-B-B delivery scenario, the agency takes on the responsibility or risk of performance with only a limited (1–2 years to primarily cover latent defects) warranty for performance resting with the builder. Agencies are now looking for ways to bundle these projects to make them attractive to a private enterprise to do most of the planning including financing the project and hence the trend leading to plan, design, construct, operate, and maintain also known as D-B-F-O-M. Tagged as Public Private Partnerships (P3) these are driven by financial strategies that take into account the risks of meeting long-term performance and hand back requirements specified by the procuring agency. Long-term Operations and Maintenance (O & M) costs are plugged into established financial models that provide lowest Net Present Values (NPV) that is used to select the most optimal (Best Value) offer to design, construct, operate and maintain infrastructure asset's on behalf of the procuring agency. Procurement agreements lay out the terms and conditions of performance in great detail along with financial disincentives if they are not met during the contract period. Hence anyone involved with this type of procurement will need to have a clear understanding of the performance criteria/requirements for the various elements and be able to offer (Technical and Financial Proposal) to design, construct, operate, and maintain in a very cost effective manner to be successful.

P3 projects are typically designed and constructed using design-build (D-B) delivery followed by the extended operate and maintain (O & M) period. Given that the private enterprise has the obligation to maintain and operate these facilities for extended periods of time ranging from 20 to 99 years, the trend is to focus the contractual requirements on performance than on "means and methods" as has been done for a long time under the traditional D-B-B method of delivery. Hence there is a need to understand performance by all those involved with this type of project delivery to be able to not only develop "performance requirements", but also understand how to deliver them. The overall performance requirements in general are all greatly influenced by geotechnical performance and hence play a very critical role in the success of any major infrastructure project.

3 Performance Requirements

A clear understanding of the purpose and need for the infrastructure will enable in the development of the procurement documents that will articulate performance criteria/requirements that are readily identified and easy to measure including a more realistic handback requirements. Hence performance criteria/requirement will need to be limited to only those elements that are easy to quantify.

Table 1 Typical geotechnical performance criteria/requirements for transportation projects

Element	Performance criteria/ requirements	Remarks
Embankments	Wood chips/shredded tires not permitted	There may be other materials that may not be permitted
Unreinforced embankment	Side slopes no steeper than 1v:3 h	Reinforced slopes may include certain restrictions as well
Foundations	Timber Piles/Battered Drilled Shafts not permitted	There may other requirements with reference to level & plumb, down drag loading, etc.
Settlement	Total (Immediate & Consolidation) < 1 in.	
	Differential < 1/2 in or 1/300	
Stability	Static F.S. > 2.0	
	Seismic F.S. > 1.1	
Liquefaction/ sinkholes etc.	Serviceability w/minimal retrofit	Others such as frost, vibration, impact to adjacent structures, etc., may be included
Erosion	Surface—Minimize	
	Sub surface—No stability concerns	

Typical Geotechnical performance criteria/requirements for infrastructure project such as a highway will include those listed below in Table 1.

Though these requirements can be spelled out based on our understanding of the functionality of the facility, there may be unique situations based on local conditions that dictate relaxing the requirements or prescribing "means and methods" to assure performance outcomes. For example, settlement of a roadway embankment may be allowed to settle greater than one inch (2.5 cm) so long as the differential settlement at a structure does not exceed one half inch (1.25 cm) or cause a sudden bump to the riding motorist at that particular interface. Hence there is a need to not only understand the requirement, but also their relevance and applicability as well.

The size and complexity of these projects are beginning to attract professionals from around the world (no longer limited to local experts) who may or may not have the local experience or understanding of the issues. This poses additional challenge to being able to design to meet the performance requirement of the project under unique local conditions.

4 Evaluation of Performance Requirements

Evaluation of performance requirements have come in the form of experiences of the geotechnical professional. Hence an experienced geotechnical professional was, is and will be critical to the success of a project. The field of geotechnical engineering has evolved over the years and the ability for acquiring data has been greatly enhanced with the advent of tools and technology. Geo professionals today have access to a lot more information than some of the pioneers in the field. However, the risk of running into unforeseeable condition is still relevant and continues to pose challenges to the geotechnical engineer. Local knowledge and experience continues to play a key role in good engineering solutions. Standard means of handling the various environmental and geotechnical conditions in a local area or a region or a country is still based on collective experience of the local professionals.

Over the years geotechnical professionals with the support of construction industry have developed "means and methods" to meet the design intent including development of tests and test procedures to verify. The test results have in the past been an indirect means to assure a certain level of desired performance with the conservatism built in the design or the process being still an unknown. With the trend of moving away from prescribing "means and methods" to allow for innovation and to hold only to performance requirements, requires a paradigm shift. In addition to making the paradigm shift one also needs to ascertain true performance over the course of a project or a life time, only through careful planning, programming, instrumentation, data collection and analysis.

Current state of practice relies heavily on individual's expertise and experience. However if monitoring for performance becomes an integral part of every program specifically when millions of dollars are being invested, the knowledge will be very valuable in understanding true performance and developing global performance standards. Advancements in monitoring tools along with ability to monitor performance on a continuous basis over extended periods even in remote locations using wireless technology coupled with cloud storage capabilities have provided a platform for easy access by our professionals. This information if mined properly and analyze by experienced individuals will provide valuable information to those developing true performance standards.

5 Developing Global Performance Standards

Performance requirements are typically, facility and element specific but are not location or environment specific. This could leave the professional including the geo professional, with a false sense of confidence in one's ability to design elements of a facility to meet the criteria without the benefit of the understanding of performance under local conditions. The private sector developer (Equity) and the builder take on the responsibility of meeting the performance criteria/requirements which are spelled out in the Project Agreement (PA) and the Inter Agreement between the Equity, Builder and the Operator & Maintainer. The design professional including the geotechnical engineer needs to be aware of the various agreements and flow down conditions while collaborating to develop a cost effective the design.

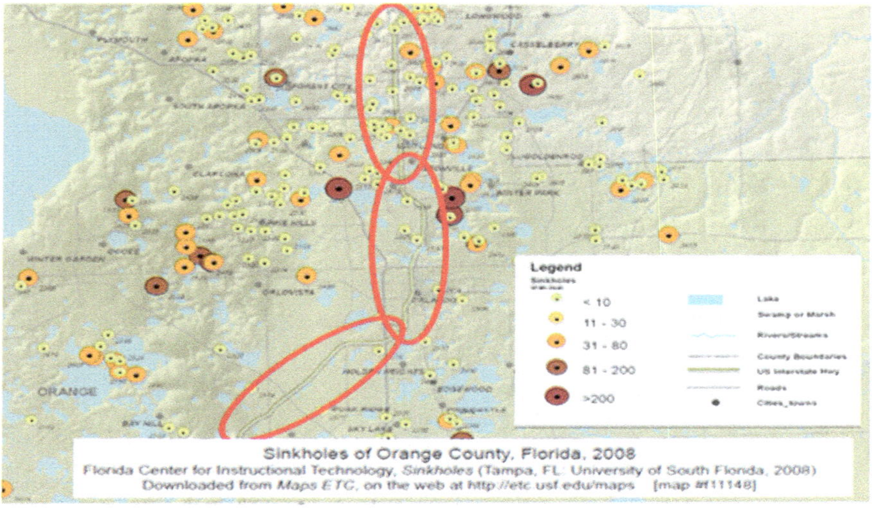

Fig. 1 Location Map—Sinkhole potential

Risk mitigation for meeting the performance and hand back requirements at the end of the contract term may be agreed to collectively by the team. This may include taking on some risk on the performance over O & M period with a prescribed certain number of maintenance intervention(s). But if the elements do not perform as anticipated, there will be an issue of who is liable for the fix or the additional maintenance. Another option to outlining performance when there is a major concern of an extreme event occurring in the region during the life of the facility and when there is no precedence for the agency; it may decide to spell out the requirement that may sound very prescriptive. An example being when Karst material forms the foundation of the region (Fig. 1) that is susceptible to development of sinkholes, the agency required the proposers to design for such an extreme event and show proof that they have done so. Typically if a sinkhole develops within roadway embankments it can be mitigated with minimal time to investigate and develop a mitigation plan that includes jet grouting. But if the sinkhole occurred under a newly constructed multi-level interchange with expensive bridge structures, the time to investigate if the bridge structures have survived the extreme event and will they be serviceable with minimal effort without endangering the traveling public is a major risk and responsibility could take months. As there was no precedence for outlining performance under these conditions, the agency decided to spell out the requirement as follows:

- Design **<u>all new foundations</u>** for new bridges to withstand the Extreme Event II Load Combination Limit State as follows:

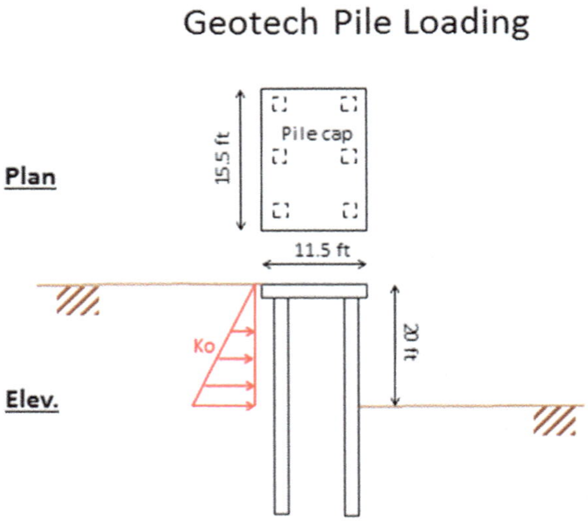

Fig. 2 Pile loading due to Sinkhole development

- Pile Groups: **Assume the ground surface on three sides of the footing cap is at proposed finished grade elevation and the ground surface under the footing cap and on one side of the footing cap is 20 ft below the proposed finished grade elevation** (Fig. 2). For piers founded on more than one pile group, assume that 20 foot subsidence occurs along the same side of all footings.
- Non-redundant Drilled Shafts: Assume the **ground surface on one side of the non-redundant drilled shaft foundation** is at proposed finished grade elevation and the ground surface on the other side of the non-redundant drilled shaft foundation is 20 ft below the proposed finished grade elevation.
- Pile/Shaft Bents: Assume the **ground surface on one side of the pile/shaft bent is at proposed finished grade elevation** and the ground surface on the other side of the pile/shaft bent is 20 ft below the proposed finished grade elevation.
- Design all new foundations for **widened bridges** to withstand the Extreme Event II Load Combination Limit State with the ground surface on only one side of the foundation **no shallower than the depth the current foundation withstands**.

This being prescriptive, the agency takes on the risk for performance but with long-term assessment, the agency hopes to develop global performance standard that will address the intent and outcome.

6 Conclusion

Can we develop true global performance requirements or standards? Yes, we can. But the key to successfully articulating the performance requirements and being able to deliver the performance are dependent on the ability of the individuals to understand behavior under a given set of conditions and predicting behavior overtime with relative confidence. However the ability to meet the global performance requirements will need to evaluated on a more regional or local basis as the practices vary from one location to another or from one region to another.

Stress State Variables for Unsaturated Soils—Consensus and Controversy

E. C. Leong

1 Introduction

Soil mechanics have successfully applied continuum mechanics to describe the response of soil to external stimuli of force and displacement. The physical difference between a soil and a continuum is obvious. A soil is a multiphase system. In a saturated soil, the two phases are the soil solid and the fluid usually water. In an unsaturated soil, three phases exist: soil solids, water and air. Although it has been put forth that a fourth phase, the air–water interface termed as the contractile skin, exists (Fredlund and Morgenstern 1977), the volume of the contractile skin is small. In terms of volume–mass relations for an unsaturated soil, the contractile skin can be neglected and its mass can be considered as part of the mass of water (Fredlund and Rahardjo 1993; Fredlund et al. 2012). Although the concept of the fourth phase seems intuitive for stress analysis of an unsaturated soil, no explicit consideration of the fourth phase in constitutive equations has been found.

The objective of this paper is to examine the stress state variables used to describe the behaviour of unsaturated soils. More specifically, this paper review the current understanding of stress state variables for unsaturated soils and evaluates the need for an equivalent effective stress for unsaturated soils by examining various forms of shear strength equations for unsaturated soils.

E. C. Leong (✉)
School of Civil & Environmental Engineering, Nanyang Technological University, Jurong West, Singapore
e-mail: cecleong@ntu.edu.sg

K. Ilamparuthi and R. G. Robinson (eds.), *Geotechnical Design and Practice*, Developments in Geotechnical Engineering,
https://doi.org/10.1007/978-981-13-0505-4_7

2 Effective Stress for Unsaturated Soils

The concept of effective stress for saturated soils as explained by Terzaghi (1936) is: "The stresses in any point of a section through a mass of soil can be computed from the total principal stresses, σ_1, σ_2, σ_3, which act at this point. If the voids of the soil are filled with water under a stress, u_w, the total principal stresses consist of two parts. One part, u_w, acts in the water and in the solid in every direction with equal intensity. It is called the neutral pressure. The balance $\sigma_1' = \sigma_1 - u_w$, $\sigma_2' = \sigma_2 - u_w$, $\sigma_3' = \sigma_3 - u_w$ represent an excess over the neutral stress, u_w, and has its seat exclusively in the solid phase of the soil. All the measurable effects of a change in shearing resistance are exclusively due to changes in the effective stress, σ_1', σ_2', σ_3'." The definition of effective stress by Terzaghi (1936) qualifies it as a stress state variable as defined in continuum mechanics and thermodynamic references.

A stress state variable is a nonmaterial variable required to characterise the stress condition (Fredlund and Rahardjo 1993; Fredlund et al. 2012). The effective stress as a stress state variable has proved pivotal in the development of saturated soil mechanics.

Various expressions of effective stress for unsaturated soils have been proposed (Nuth and Laloui 2008). These forms are summarised in Table 1. The form of effective stress for unsaturated soils suggested by Croney et al. (1958) is equivalent to the form suggested by Bishop if pore-air pressure is atmospheric pressure

Table 1 Effective stress for unsaturated soils

References	Effective stress	Notations
Croney et al. (1958)	$\sigma' = \sigma - \beta' u_w$	σ' = effective normal stress σ = normal stress β' = holding or bonding factor, effective in contributing to the shear strength of soil u_w = pore-water pressure
Bishop (1959)	$\sigma' = (\sigma - u_a) + \chi(u_a - u_w)$	u_a = pore-air pressure χ = a parameter related to degree of saturation of soil
Aitchison (1961)	$\sigma' = \sigma + \psi p$	p = pore-water pressure deficiency ψ = a parameter with values ranging from 0 to 1
Jennings (1961)	$\sigma' = \sigma + \beta' \lvert u_w \rvert$	β = a statistical factor of the same type as the contact area
Richards (1966)	$\sigma' = \sigma - u_a + \chi_m(h_m + u_a) + \chi_s(h_s + u_a)$	χ_m = effective stress parameter h_m = matric suction (cm of water) χ_s = effective stress parameter for solute suction h_s = solute or osmotic suction (cm of water)
Aitchison (1965, 1973)	$\sigma' = \sigma - u_a + \chi_m(u_a - u_w) + \chi_s(\pi)$	π = solute or osmotic suction χ_m and χ_s = soil parameters which are usually in the range 0–1, dependent on stress path

(i.e. $u_a = 0$ kPa). The inclusion of solute suction or osmotic suction by Richards (1966) and Aitchison (1965, 1973) warrants further explanation. Soil suction or total suction consists of mainly two components: matric suction (u_a–u_w) and osmotic suction (π) (Krahn and Fredlund 1972). Matric suction can be measured directly by measuring the negative pore-water pressure u_w using for example, a tensiometer (e.g. He et al. 2006). Osmotic suction which is attributed to the presence of dissolved salt in the pore water cannot be measured directly, it can be inferred by measuring the salt concentration of the pore-water or by measuring the total suction and matric suction (e.g. Leong et al. 2003).

The effect of matric suction on shear strength of unsaturated soils has been observed in laboratory element tests (e.g. Bishop et al. 1960; Bishop and Blight 1963; Escario and Saez 1986; Rahardjo et al. 2004; Nyunt et al. 2011). However, the effect of osmotic suction on shear strength of unsaturated soils has not been positively observed. Tang et al. (1997) observed that shear strength decreases slightly as osmotic suction increases while Katte and Blight (2012) did not observed any effect of osmotic suction on shear strength. The salt concentration in the pore-water influences the osmotic suction through its influence of the diffuse double layer around clay particles (Fredlund and Rahardjo 1993; Fredlund et al. 2012). Thyagaraj and Salini (2015) provided experimental evidence of the influence of osmotic suction on the double diffuse layer in expansive clay soils. It was concluded by Tang et al. (1997) that total suction cannot be as a stress state variable for unsaturated soils. Two reasons were provided in Fredlund and Rahardjo (1993) for not including osmotic suction in geotechnical engineering problems: (1) The total suction and matric suction curves are almost congruent. Therefore, a change in total suction is equivalent to a change in matric suction; and (2) Laboratory test results have already accounted for the changes in osmotic suction if the changes occurring in the field are simulated in the laboratory test. Hence from Table 1, the common form of effective stress that has been suggested for unsaturated soils can be taken as the form suggested by Bishop (1959).

3 Shear Strength of Unsaturated Soils

Shear strength of soils is a constitutive relation. It is a single-valued equation that expresses the relationship between state variables. In more complex models such as the Barcelona Basic Model (Alonso et al. 1990) or SFG model (Sheng et al. 2008), the model formulation maintains the separation of stress state variables and constitutive behaviour. The most likely reason for separation of stress state variables and constitutive behaviour in these models is that more than one constitutive relation is present and using two stress state variables provided more flexibility in the formulation of the constitutive relationships. The problem with the effective stress for unsaturated soils is that it needs a different form of effective stress for shear strength and volume change constitutive relations and thus it is easier to separate the stress state variables (Fredlund 2015).

The major controversy in effective stress for unsaturated soils arises especially when only one constitutive relationship is examined. Sometimes this is due to the misunderstanding of the definition of stress state variable coupled with the form of the constitutive relationship. This is illustrated by a review of the shear strength equations that have been proposed for unsaturated soils as summarised in Table 2. The shear strength equation should be valid for the full range of saturation, i.e. from dry to full saturation. All of the shear strength equations in Table 2 satisfies this condition and reverts to the shear strength equation for saturated soils, i.e.

$$\tau = c' + (\sigma - u_{\mathrm{w}}) \tan \phi' = c' + \sigma' \tan \phi' \quad (1)$$

All the shear strength equations in Fig. 2 can also be easily re-cast into the form of Bishop and Blight (1961) shear strength equation

$$\tau = c' + [(\sigma - u_{\mathrm{a}}) + \chi(u_{\mathrm{a}} - u_{\mathrm{w}})] \tan \phi', \quad (2)$$

where χ is the Bishop's effective stress parameter. The equivalent forms of χ for the shear strength equations are shown in Table 3. From Table 3, it can be observed that χ can be expressed in terms of volumetric water content, degree of saturation, matric suction and shear strength parameters. The various forms suggest that χ is not unique for different soil types. A fact that has been well established by experimental evidence is shown in Fig. 1.

4 Stress State Variables for Unsaturated Soils

Fredlund and Morgenstern (1977) put forth the theoretical basis for independent stress state variables for unsaturated soils based on a continuum mechanics approach. They further provided combinations of stress variables that can be used to define the state variables. The three possible combinations of stress state variables are $(\sigma - u_{\mathrm{a}})$, $(\sigma - u_{\mathrm{w}})$ and $(u_{\mathrm{a}} - u_{\mathrm{w}})$. Only two combinations of stress state variables are needed to describe the constitutive relations of unsaturated soils. The most convenient combinations for engineering practice are net normal stress $(\sigma - u_{\mathrm{a}})$ and matric suction $(u_{\mathrm{a}} - u_{\mathrm{w}})$.

The justification for these stress state variables has been repeatedly evaluated by others. Zhang and Lytton (2006) showed that three stress state variables, $(\sigma - u_{\mathrm{a}})$, $(u_{\mathrm{a}} - u_{\mathrm{w}})$ and u_{a} are needed if the pore-air pressure influence on volume change is not negligible. Lu (2008) demonstrated that matric suction is definitely not a stress variable but could be a stress state variable by using 'universally accepted concepts of mechanical equilibrium, stress definition on an REV, and physical and logical reasoning'. However, there is an interdependency between net normal stress and matric suction, and other stress state variables may be necessary to describe the full behaviour of unsaturated soils.

Table 2 Shear strength equations for unsaturated soils

References	Shear strength equation	Notations
Fredlund et al. (1978)	$\tau = c' + (\sigma - u_a)\tan\phi' + (u_a - u_w)\tan\phi^b$	τ = shear strength c' = effective cohesion ϕ' = effective friction angle ϕ_b = angle indicating a change in shear strength related to matric suction
Lambom (1986)	$\tau = c' + (\sigma - u_a)\tan\phi' + \theta(u_a - u_w)\tan\phi'$	θ = volumetric water content
Vanapalli et al. (1996)	$\tau = c' + (\sigma - u_a)\tan\phi' + \left(\frac{\theta - \theta_r}{\theta_s - \theta_r}\right) \times (u_a - u_w)\tan\phi'$	θ_r = residual volumetric water content θ_s = saturated volumetric water content
Oberg and Sallfors (1997)	$\tau = c' + (\sigma - u_a)\tan\phi' + S(u_a - u_w)\tan\phi'$	S = degree of saturation
Bao et al. (1998)	$\tau = c' + (\sigma - u_a)\tan\phi' + \left\{\frac{\log\left[\frac{(u_a - u_w)_r}{(u_a - u_w)}\right]}{\log\left[\frac{(u_a - u_w)_r}{(u_a - u_w)_b}\right]}\right\} \times (u_a - u_w)\tan\phi'$	$(u_a - u_w)_r$ = residual matric suction $(u_a - u_w)_b$ = air-entry value
Khalili and Khabbaz (1998)	$\tau = c' + (\sigma - u_a)\tan\phi' + \left[\frac{(u_a - u_w)}{(u_a - u_w)_b}\right]^{-0.55} \times (u_a - u_w)\tan\phi'$	
Aubeny and Lytton (2003)	$\tau = c' + (\sigma - u_a)\tan\phi' + f_1(\theta)(u_a - u_w)\tan\phi'$	$f_1(\theta) = \begin{cases} \frac{1}{\theta} \text{ for } S = 100\% \\ 1 + \frac{S - 85}{15}\left(\frac{1}{\theta} - 1\right) \text{ for } 85\% \le S \le 100\% \\ 1 \text{ for } S \le 85\% \end{cases}$

(continued)

Table 2 (continued)

References	Shear strength equation	Notations
Tekinsoy et al. (2004)	$\tau = c' + (\sigma - u_a)\tan\phi' + [(u_a - u_w) + P_a] \times \ln\left[\frac{(u_a - u_w) + P_a}{P_a}\right]\tan\phi$	P_a = atmospheric pressure
Garven and Vanapalli (2006)	$\tau = c' + (\sigma - u_a)\tan\phi' + \left(\frac{\theta - \theta_r}{\theta_s - \theta_r}\right)^{\kappa}(u_a - u_w)\tan\phi'$	$\kappa = -0.0016I_p^2 + 0.0975I_p + 1$
Vilar (2006)	$\tau = c' + (\sigma - u_a)\tan\phi' + \left[\frac{(c_{ult} - c')}{(c_{ult} - c') + (u_a - u_w)\tan\phi'}\right] \times (u_a - u_w)\tan\phi'$	c_{ult} = ultimate shear strength

Table 3 Equivalent χ for shear strength equations listed in Table 2

References	$\chi =$
Fredlund et al. (1978)	$\frac{\tan \phi^{b}}{\tan \phi'}$
Lambom (1986)	θ
Vanapalli et al. (1996)	$\left(\frac{\theta-\theta_r}{\theta_s-\theta_r}\right)$
Oberg and Sallfors (1997)	S
Bao et al. (1998)	$\left\{\frac{\log\left[\frac{(u_a-u_w)_r}{(u_a-u_w)}\right]}{\log\left[\frac{(u_a-u_w)_r}{(u_a-u_w)_b}\right]}\right\}$
Khalili and Khabbaz (1998)	$\left[\frac{(u_a-u_w)}{(u_a-u_w)_b}\right]^{-0.55}$
Aubeny and Lytton (2003)	$f_1(\theta)$
Tekinsoy et al. (2004)	$\frac{[(u_a-u_w)+P_a]}{(u_a-u_w)} \ln\left[\frac{(u_a-u_w)+P_a}{P_a}\right]$
Garven and Vanapalli (2006)	$\left(\frac{\theta-\theta_r}{\theta_s-\theta_r}\right)^{\kappa}$
Vilar (2006)	$\left[\frac{(c_{ult}-c')}{(c_{ult}-c')+(u_a-u_w)\tan\phi'}\right]$

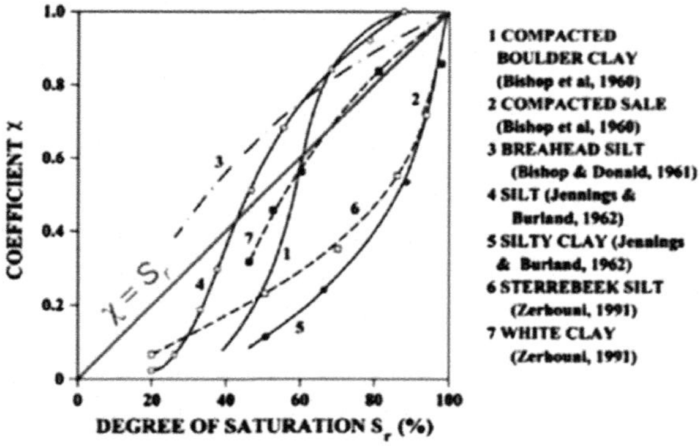

Fig. 1 Effective stress parameter χ versus degree of saturation (from Zerhouni 1991)

The general consensus is that the stress state variables of net normal stress (σ–u_a) and matric suction (u_a–u_w) are needed to describe the behaviour of unsaturated soils. However, additional stress state variables are needed to describe the complete constitutive behaviour of unsaturated soils. The roles of these additional stress state variables have not been clearly demonstrated through experiments or observations in the field.

Currently, the most contentious stress state variable is matric suction (u_a–u_w). Matric suction is due to the pore-water that exists in meniscus form between soil particles. As a soil dries, the pore-water may exists in insular air saturation, fuzzy saturation and pendular saturation as illustrated in Fig. 2 (Kohgo et al. 1993). Considering that the arrangement of the pore-water and the associated menisci in an unsaturated soil is dependent on the particle shape, size, arrangement, pore-water chemistry and particle surface properties, the effect of matric suction on constitutive

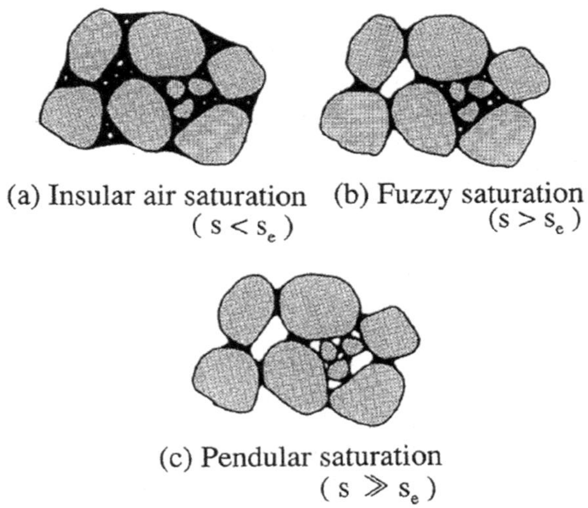

Fig. 2 Possible saturation conditions in real soils (from Kohgo et al. 1993)

behaviour of soil is non-uniform and non-homogeneous. There have been attempts to account for the non-uniform and non-homogeneous action of matric suction through the use of a suction stress as a stress state variable instead of matric suction (e.g. Karube et al. 1996; Lu and Likos 2006). The use of suction stress has been shown to work well for shear strength behaviour of unsaturated soils (e.g. Lu and Likos 2006; Kim et al. 2010) but it remains to be fully validated for other constitutive relationships.

5 Conclusion

In the study of unsaturated soil mechanics, attempts have been made to use a single stress state variable to describe the constitutive relationships of soils. Limited success of using a single stress state variable was achieved with some soils. Generally, it is more well accepted that two stress state variables, net normal stress and matric suction, be used to describe the constitutive behaviour of unsaturated soils. Many element tests have been conducted where these two stress state variables were applied or measured. More advanced constitutive models have been developed using these two stress state variables and have been demonstrated to be able to describe more complete unsaturated soil behaviour. In recent years, it has been suggested that more than two stress state variables may be needed to describe all constitutive behaviour of unsaturated soils. It has also been suggested that the stress state variable, matric suction, be replaced with suction stress instead. More study and research are needed to investigate these suggestions.

References

Aitchison, G.D. 1961. Relationship of moisture and effective stress functions in unsaturated soils. In *Pore Pressure and Suction in Soils Conference, organised by British Natural Society of International. Society of Soil Mechanics and Foundation Engineering*. At Institution Civil Engineering, 47–52. London, England: Butterworth.

Aitchison, G.D. 1965. Soil properties, shear strength, and consolidation. In *Proceedings of 6th International Conference on Soil Mechanics and Foundation Engineering, Montreal, Canada*, vol. 3, 318–321.

Aitchison, G.D. 1973. The quantitative description of the stress-deformation behaviour of expansive soils-Preface to the set of papers. In *Proceedings of 3rd International Conference on Expansive Soils, Haifa, Isreal*, vol. 2, 83–88.

Alonso, E.E., A. Gens, and A. Josa. 1990. A constitutive model for partially saturated soils. *Geotechnique* 40 (3): 405–430.

Aubeny, C., and R. Lytton. 2003. *Estimating strength versus location and time in high plasticity clays*. College Station, Texas: Texas Transportation Institute.

Bao, C.G., B. Gong, and L. Zhan. 1998. Properties of unsaturated soils and slope stability of expansive soils. In *Proceedings of the 2nd International Conference on Unsaturated Soils (UNSAT 98)*, vol. 1, 71–98.

Bishop, A.W. 1959. The principle of effective stress. *Teknisk Ukeblad* 106 (39): 859–863. (Norwegian Geotechnical Institute).

Bishop, A.W., I. Alpan, G.E. Blight, and I.B. Donald. 1960. Factors controlling the shear strength of partially saturated cohesive soils. In *ASCE Research Conferences on Shear Strength of Cohesive Soils*, 503–532. Boulder: University of Colorado.

Bishop, A.W., and G.E. Blight. 1963. Some aspects of effective stress in saturated and partly saturated soils. *Geotechnique* 13: 177–197.

Croney, D., J.D. Coleman, and W.P.M. Black. 1958. Movement and distribution of water in relation to highway design and performance. In *Water and Its Conduction in Soils, Highway Research Board, Special Report*, no. 40, 226–252. Washington, DC.

Escario, V., and J. Saez. 1986. The strength of partly saturated soils. *Geotechnique* 36 (3): 453–456.

Fredlund, D.G., and H. Rahardjo. 1993. *Soil Mechanics for Unsaturated Soils*, 507p. New York: Wiley.

Fredlund, D.G., H. Rahardjo, and M.D. Fredlund. 2012. *Unsaturated Soil Mechanics in Engineering Practice*, 926p. New York: Wiley.

Fredlund, D.G. 2015. State variables in saturated-unsaturated soil mechanics. *Soils and Rocks* 39 (1): 3–17.

Fredlund, D.G., and N.R. Morgenstern. 1977. Stress state variables for unsaturated soils. *Journal of Geotechnical Engineering Division, ASCE* 103 (GT5): 447–466.

Fredlund, D.G., N.R. Morgenstern, and R.A. Widger. 1978. Shear strength of unsaturated soils. *Canadian Geotechnical Journal* 15 (3): 313–321.

Garven, E. A., and S.K. Vanapalli. 2006. Evaluation of empirical procedures for predicting the shear strength of unsaturated soils. In *Proceedings of 4th International Conference of Unsaturated Soil, UNSAT 2006, ASCE Geotechnical Special Publication 147, ASCE,* Reston, Va., 2570–2581.

He, L., E.C. Leong, and A. Algamal. 2006. A miniature tensiometer for measurement of high matric suction, In *Proceedings of 4th International Conference on Unsaturated Soils*, ed. G.A. Miller, C.E. Zapata, S.L. Houston, & D.G. Fredlund, Carefree, Arizona, April 2–6, 2006: 1897–1907. United States: American Society of Civil Engineers.

Jennings, J.E. 1961. A revised effective stress law for use in prediction of the behaviour of unsaturated soils. In *Proceedings of Conference on Pore Pressure and Suction in Soils*, 26–30. London, England.

Karube, D., S. Kato, K. Hamada, and M. Honda. 1996. The relationship between the mechanical behavior and the state of pore water in unsaturated soil. *Geotechnical Engineering Journal, JSCE*, 535 (III-34): 83–92.
Katte, V.Y., and G.E. Blight. 2012. The roles of solute suction and surface tension in the strength of unsaturated soil. In *Unsaturated Soils: Research and Applications*, vol. 2, ed. C. Mancuso, C. Jommi, and F. D'Onza, 431–438. Springer: Heidelberg, Germany.
Khalili, N., and M.H. Khabbaz. 1998. A unique relationship for the determination of the shear strength of unsaturated soils. *Geotechnique* 48 (5): 681–687.
Kim, B.-S., S. Shibuya, S.-W. Park, and S. Kato. 2010. Application of suction stress for estimating unsaturated shear strength of soils using direct shear testing under low confining pressure. *Canadian Geotechnical Journal* 47: 955–970.
Kohgo, Y., M. Nakano, and T. Miyazaki. 1993. Theoretical aspects of constitutive modelling for unsaturated soils. *Soils and Foundations* 33 (4): 49–63.
Krahn, J., and D.G. Fredlund. 1972. On total, matric and osmotic suction. *Journal of Soil Science* 114 (5): 339–348.
Lamborn, M.J. 1986. *A Micromechanical Approach to Modelling Partly Saturated Soils*. M.Sc. Thesis, Texas A&M University, Texas.
Leong, E.C., S. Tripathy, and H. Rahardjo. 2003. Total suction measurement of unsaturated soils with a device using the chilled-mirror dewpoint technique. *Geotechnique* 53 (2): 173–182.
Lu, N. 2008. Is matric suction a stress variable? *Journal of Geotechnical and Geoenvironmental Engineering* 134 (7): 899–905.
Lu, N., and W.J. Likos. 2006. Suction stress characteristic curve for unsaturated soil. *Journal of Geotechnical and Geoenvironmental Engineering* 132 (2): 131–142.
Nuth, M., and L. Laloui. 2008. Effective stress concept in unsaturated soils: Clarification and validation of a unified framework. *International Journal For Numerical and Analytical Methods in Geomechanics* 32: 771–801.
Nyunt, T.T., E.C. Leong, and H. Rahardjo. 2011. Strength and small-strain stiffness characteristics of unsaturated sand. *Geotechnical Testing Journal* 34 (5): 551–561.
Oberg, A., and G. Sallfors. 1997. Determination of shear strength parameters of unsaturated silts and sands based on water retention curve. *Geotechnical Testing Journal* 20: 40–48.
Rahardjo, H., Ong Boo Heng, and E.C. Leong. 2004. Shear strength of a compacted residual soil from consolidated drained and constant water content triaxial tests. *Canadian Geotechnical* 41 (3): 421–436.
Richards, B.G. 1966. The significance of moisture flow and equilibria in unsaturated soils in relation to the design of engineering structures built on shallow foundations in Australia. Presented at the Symposium On Permeability and Capillary, American Society Testing Materials, Atlantic City, N.J.
Sheng, D., D.G. Fredlund, and A. Gens. 2008. A new modelling approach for unsaturated soils using independent stress variables. *Canadian Geotechnical Journal* 45 (5): 511–534.
Tang, G.X., J. Graham, and D.G. Fredlund. 1997. Effect of osmotic suction on strength of unsaturated highly plastic clays. In *Proceedings of 50th Canadian Geotechnical Conference*, vol. 2, pp. 641–648. Golden Jubilee, Ottawa, Canada, October 20–27.
Tekinsoy, M.A., C. Kayadelan, M.S. Keskin, and M. Soylemaz. 2004. An equation for predicting shear strength envelope with respect to matric suction. *Computers and Geotechnics* 31 (7): 589–593.
Terzaghi, K. 1936. The shearing resistance of saturated soils. In *Proceedings 1st International Conference on Soil Mechanics*, vol. 1.
Thyagaraj, T., and U. Salini. 2015. Effect of pore fluid osmotic suction on matric and total suctions of compacted clay. *Géotechnique* 65 (11): 952–960.
Vanapalli, S.K., D.G. Fredlund, D.E. Pufahl, and A.W. Clifton. 1996. Model for the prediction of shear strength with respect to soil suction. *Canadian Geotechnical Journal* 33: 379–392.
Vilar, O.M. 2006. A simplified procedure to estimate the shear strength envelope of unsaturated soil. *Canadian Geotechnical Journal* 43: 1088–1095.

Zerhouni, M.I. 1991. Role de la pression interstitielle negative dans le comportement des sols—apllication au calcul des routes. Ph.D. Thesis, Ecole Centrale Paris, Paris, France.
Zhang, X., and R. Lytton. 2006. Stress state variables for saturated and unsaturated Soils. *Unsaturated Soils* 2006: 2380–2391.

Addressing Transportation Geotechnics Challenges Using Mechanics of Unsaturated Soils

Sai K. Vanapalli and Zhong Han

1 Introduction

The present-day highway and local road networks are expected to facilitate safe and cost effective transportation of products as well as the passengers. The industrial, business and employment opportunities that significantly contribute to the economic development of a region or a nation are closely dependent on the land transportation system facilities. In addition, high-quality transportation system significantly reduces the associated maintenance and rehabilitation costs. For example, as of 2015, India has a road network of over 5,472,144 km; Canada has a national highway system of 38,021 km and the similar system in the USA is 259,032 km (data obtained from Wikipedia https://en.wikipedia.org). The maintenance and rehabilitation costs of transportation facilities can be huge, if the transportation facilities are not of high quality. These costs can have both a direct and an indirect effect on the economic development. Even a small percentage of reduction in the associated rehabilitation costs will result in significant savings.

One of the major geotechnics challenge for the transportation systems is the significant influence of the external environment (seasonal moisture and temperature changes and the associated drying–wetting and freeze–thaw cycles) on the pavement performance and service life. If the influence of the environmental factors is not properly addressed in the design of new and rehabilitated pavements; there will be a high likelihood of unsatisfactory serviceability, exorbitant maintenance costs, and

S. K. Vanapalli (✉) · Z. Han
Department of Civil Engineering, University of Ottawa, Ottawa K1N 6N5, Canada
e-mail: sai.vanapalli@uottawa.ca

Z. Han
e-mail: nicholas.han@hotmail.com

S. K. Vanapalli · Z. Han
School of Civil Engineering, Wuhan University, Wuhan 430072, China

K. Ilamparuthi and R. G. Robinson (eds.), *Geotechnical Design and Practice*, Developments in Geotechnical Engineering,
https://doi.org/10.1007/978-981-13-0505-4_8

reduced life span. Pavement is a layered structure of compacted materials maintained with various measures to be in a state of unsaturated condition throughout their design life. In recent years, the influence of the moisture and temperature on the mechanical properties of the pavement materials has been successfully interpreted and predicted within the framework of the mechanics of unsaturated soils using suction as the key stress state variable (Khoury and Zaman 2004; Zapata et al. 2007; Ng and Zhou 2014; Han and Vanapalli 2015; Li et al. 2015).

This paper presents some recent advancements in the mechanics of unsaturated soils and pavement design theory that can be used to address the influence of the environmental factors on the performance of pavements. The soil–water characteristic curve (SWCC) and the M_R are the key material properties required for performing hydro-mechanical analysis for pavement structure. Several approaches proposed for predicting the SWCC and the variation of the resilient modulus (M_R) with moisture content and suction for different fine-grained soils using limited and easy-to-obtain experimental data are introduced and integrated. These approaches can be used in pavement engineering practice for addressing the transportation geotechnics challenges taking account of the influence of the various environmental factors. The proposed integrated approach is promising and more importantly, simple and economical and can be used for rational design of pavements based on state-of-the-art understanding of the mechanics of unsaturated soils.

2 Influence of Environmental Factors on Pavement Design

2.1 Pavement Design Theory

Pavements are layered structures. As shown in Fig. 1, a typical pavement structure is formed with compacted granular materials as base and sub-base layers, which are placed over compacted subgrade layer and sealed with flexible and/or rigid surface layer.

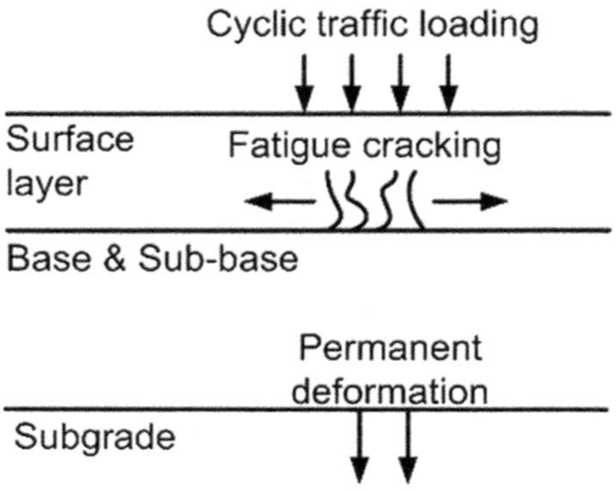

Fig. 1 Typical pavement structure and failure criteria

Unlike the design theory for shallow and deep foundations which are based on limiting shear failure/deflection, the two major failure criteria for pavements are the fatigue cracking that initiates at the bottom of the surfacing layer and the rutting that happens at the surface of the subgrade layer (see in Fig. 1). The life span of a pavement structure in the state-of-the-art mechanistic pavement design methods is calculated based on its serviceability that is determined by the fatigue cracking and rutting.

The fatigue cracking is related to the local tensile strain and the elastic modulus of the surfacing materials. For example, Eq. (1) is used to determine the allowable number of load repetitions (N_f) for flexible pavements, which directly determines the life span of the pavement, based on the fatigue cracking criterion (Huang 2004).

$$N_f = f_1(\varepsilon_t)^{-f_2}(E_1)^{-f_3}, \quad (1)$$

where ε_t is the tensile strain at the bottom of the asphalt layer, E_1 is the elastic modulus of the asphalt surfacing and the f_1, f_2, and f_3 are material constants. The ε_t is due to the formation of the deflection basin under wheel loads and is calculated using elastic theory by assuming that the base, sub-base and subgrade layers are essentially elastic.

The resilient modulus (M_R) is defined as the ratio of the cyclic deviator stress (σ_d which is to simulate the cyclic traffic loading) to the resulted resilient strain (ε_r) in the pavement materials. The M_R represents the material stiffness under cyclic loading and is used as the elastic modulus of pavement materials. Due to this reason, the M_R is considered to be the key material property required in the mechanistic pavement design methods to characterize the resilient behavior of pavement materials, dimension the multilayer system of a pavement structure and analyze the fatigue cracking typically initiated at the bottom of the asphalt or concrete surface layers (Han and Vanapalli 2015).

2.2 *Influence of External Environment on the Resilient Modulus*

The moisture content and temperature of compacted pavement materials fluctuate due to the influence of various seasonal environmental factors such as the infiltration, evaporation, freeze–thaw cycles and ground water table variation. It is well known that the mechanical behavior of compacted subgrade of the pavement varies significantly with moisture content and temperature. Figure 2 shows typical evolution of pavement capacity, which is mainly determined by the M_R of pavement layers, during a period of one year taking account of the influence of different environmental factors.

It can be seen from Fig. 2 that soil starts to freeze in late fall (e.g., November) and remains in a frozen state during winter (e.g., December to March, Line AB

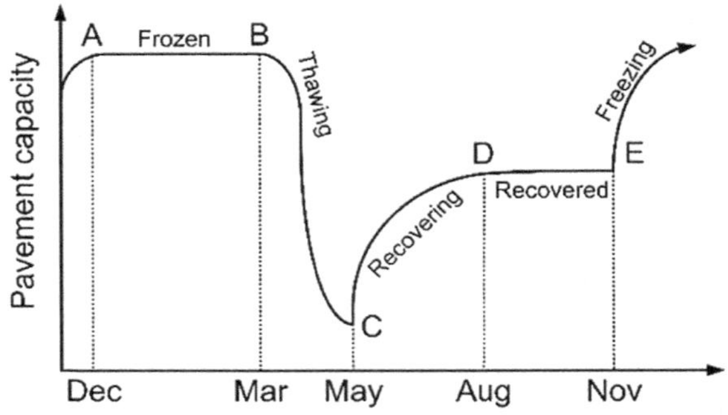

Fig. 2 Variation in pavement stiffness under freeze–thaw condition (modified after Jong et al. 1998)

in Fig. 2). The M_R of pavement materials after freezing typically increases 20–120 times (Bosscher and Nelson 1987; Bigl and Berg 1996; Zapata et al. 2007).

The pavement capacity and M_R reduce dramatically during the spring thaw period (i.e. March to May, Line BC in Fig. 2). Several reasons can be attributed to this decrease;

- The volume of soil during freezing increases upon the formation of the ice lenses, which decreases the soil's dry density and cohesion and increases its moisture content. Such processes weaken the soil's structure;
- The moisture content of the soil will increase during the thawing process.

Several studies have shown that the M_R after thawing can be 50–40% less than the M_R of the same soil that is never frozen (Lee et al. 1995; Jong et al. 1998; Li et al. 2015). Pavement capacity and M_R gradually recover during the summer and fall period (i.e. May to November, Line CD and Line DE in Fig. 2) due to the drainage and evaporation that lead to decrease in the soil moisture content.

2.3 Pavement Design Using the Mechanics of Unsaturated Soils

Both the temperature and the moisture content influence the M_R. For this reason, the pavement performance in different seasons should be well understood. Typically, M_R increases with a decrease in the moisture content (LeKarp et al. 2000; Khoury et al. 2013; Li et al. 2015; Han and Vanapalli 2016a). In order to investigate the influence of moisture content on the pavement behavior, the following two pieces of information are required:

- Moisture content distribution and changes associated with various environmental factors within the pavement layers;
- The resulting changes in the M_R of pavement materials.

As-compacted soils are typical examples of unsaturated soils, the hydro-mechanical behavior of which should be interpreted within the framework of the mechanics of unsaturated soils using suction (s) as the key stress state variable. The soil–water characteristic curve (SWCC) defines the relationship between the moisture content and suction. The SWCC along with the permeability function forms the fundamental constitutive relationship for the hydraulic analysis of unsaturated soils (Fredlund 2006). The SWCC has been successfully used for analyzing the moisture migration and distribution within pavement layers using numerical methods (Zapata et al. 2007).

In addition, the variation in the M_R with moisture content can be interpreted and predicted by establishing the M_R–s relationships (Sawangsuriya et al. 2009; Cary and Zapata 2011; Salour et al. 2014; Coronado et al. 2016). Detailed review and discussion on determining and predicting the M_R–s relationships are available in Han and Vanapalli (2016a). From these studies, it can be understood that the SWCC and the M_R–s or M_R–moisture content relationships are the key information needed for the reliable pavement drainage and structural design and analysis.

3 Integrated Approach for Predicting the SWCC and the M_R

3.1 Predicting the M_R–Moisture Content Relationships

Pavement materials are typically compacted at optimum moisture content (w_{opt}) to achieve maximum dry density (ρ_{dmax}) and higher stiffness and shear strength properties (Khoury and Zaman 2004). The M_R at optimum moisture content (M_{Ropt}) of pavement materials is conventionally determined and used in the design of pavements.

Han and Vanapalli (2015) developed Eq. (2) to relate the M_R to the suction (s) and gravimetric water content (w) using corresponding values at optimum moisture content (indicated using subscript opt) and saturated moisture content (indicated using subscript sat) and one model parameter ξ.

$$\frac{M_R - M_{Rsat}}{M_{Ropt} - M_{Rsat}} = \frac{s}{s_{opt}} \left(\frac{w}{w_{opt}} \right)^{\xi} \tag{2}$$

The following advantages can be highlighted for using Eq. (2) for predicting the M_R–s or M_R–w relationships:

(i) The w corresponding to the s or the s corresponding to the w can be estimated using the SWCC;

(ii) Equation (2) was validated by Han and Vanapalli (2015) using data derived from 11 different compacted subgrade soils. Han and Vanapalli (2015) suggested that the model parameter ξ, which should be determined from regression analysis performed on a large number of testing data at various s levels, typically falls within a narrow range between 1.0 and 3.0. An intermediate ξ value of 2.0 provides reasonable predictions for different subgrade soils. If a constant ξ value of 2.0 is used, the required information in the model for prediction is reduced only to the SWCC, M_{Rsat}, M_{Ropt}, and s_{opt} (for predicting M_R–s relationship) or w_{opt} (for predicting M_R–w relationship). The experimental data on the M_R at suction levels other than s_{opt} and $s = 0$ are required by many other prediction models in the literature. However, these testing data are not required for using Eq. (2). In other words, Eq. (2) significantly simplifies the prediction procedure;

(iii) The M_{Rsat} and M_{Ropt}, which are required in Eq. (2), can be measured using conventional cyclic loading tests without suction control. They are also the key information required in the mechanistic pavement design methods to analyze the as-compacted pavement layers (using M_{Ropt}) and the layers under the worst scenario (using M_{Rsat} for layers that are totally saturated);

(iv) The influence of the external stress can also be taken into account by using relevant information of M_{Rsat} and M_{Ropt} in Eq. (2).

3.2 *Predicting the SWCC*

Experimental data of the s_{opt}, w_{opt} and the SWCC are required in Eq. (2). There are several approaches available in the literature for predicting the SWCC of fine-grained soils that require a SWCC family and one measurement on the SWCC (e.g., Catana et al. 2006; Houston et al. 2006; Chin et al. 2010). The principle of these approaches is illustrated in Fig. 3.

Perera et al. (2005) suggested a SWCC family based on Fredlund and Xing (1994) equation (Eq. 3) for fine-grained soils that is differentiated by the wPI (i.e., wPI = I_p (%silt + %clay)/100 where I_p is the plasticity index of the soil) as shown

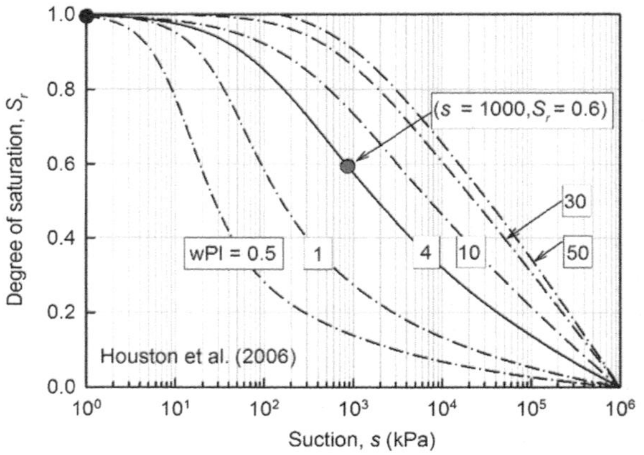

Fig. 3 Principle of Houston et al. (2006) method for predicting the SWCC

in dashed lines in Fig. 3. The wPI is related to the model parameters a, n, m and s_{res} in Eq. (3) through Eq. (4).

$$S_r = \frac{w}{w_{sat}} = \left[1 - \frac{\ln(1 + s/s_{res})}{\ln(1 + 10^6/s_{res})}\right] \frac{1}{\{\ln[2.718 + (s/a)^n]\}^m} \quad (3)$$

$$\begin{cases} a = 32.835\ \ln(\mathrm{wPI}) + 32.438 \\ n = 1.421\ \mathrm{wPI}^{-0.3185} \\ m = -0.2154\ \ln(\mathrm{wPI}) + 0.7145 \\ s_{res} = 500 \end{cases} \quad (4)$$

Houston et al. (2006) suggested selecting the predicted SWCC using one measurement of s–moisture content relationship from the SWCC family described by Eq. (4). For example, if one measurement (say $S_r = 0.6$ at $s = 1000$ kPa as shown in Fig. 3) is available for a soil, the SWCC passing through this measurement, which is shown as a solid line, is the predicted SWCC for this soil. The wPI value corresponding to the SWCC passing through the measured data point at an unsaturated condition can be obtained by substituting Eq. (4) and the measured point (e.g., $s = 1000$ kPa, $S_r = 0.6$) into Eq. (3). For the case shown in Fig. 3, the predicted SWCC is described by Eq. (3) using a wPI value of 4.

The measured s_{opt}–w_{opt} (i.e., one data set of measurement) is required in Eq. (2) for predicting the SWCC in Houston et al. (2006) method. The approaches introduced in Sects. 3.1 and 3.2 for predicting the SWCC and the M_R, respectively are integrated. Both these methods need measurements at the same moisture conditions and the predicted SWCC is used in Eq. (2) for predicting the M_R. Such an integrated approach would reduce the information required for predicting the M_R–moisture content relationship and the SWCC of pavement fine-grained subgrade soils to: M_{Rsat}, M_{Ropt}, s_{opt}, w_{opt}, and w_{sat}.

4 Experimental Validation of the Model

A comprehensive experimental program sponsored by the Ministry of Transportation of Ontario, Canada has been conducted to determine the (i) M_R and its variation with moisture content and suction and (ii) the SWCC for typical Ontario subgrade soils. Experimental data from this project on three soils (i.e., Kincardine lean clay, KLC; Toronto silty clay, TSC; Ottawa lean clay, OLC) are shown and analyzed in this paper to validate the approaches introduced. The various soil properties of three soils are summarized in Table 1.

The specimens were initially compacted at w_{opt} and then brought to different higher or lower moisture contents before performing the cyclic loading tests as per AASHTO T307-99 testing protocol. Suction levels of the specimens after the cyclic loading tests and the SWCC of the specimens were measured using filter paper methods. Detailed experimental procedures are available in Han and Vanapalli (2016b).

Table 1 Physical properties of five Ontario subgrade soils

Soil ID	KLC	OLC	TSC
w_L (%)	31	48	19.6
I_p	10	26	6
w_{opt} (%)	20.3	23.0	13.5
S_{ropt} (%)	84	90	90
γ_{dmax} (kN/m^3)	16.31	16.16	19.15
G_s	2.71	2.75	2.68
%sand	15	20	3
%silt	60	48	81
%clay	25	32	16
AASHTO	A-4	A-6	A-4
USCS	CL	CL	CL-ML

Note w_L = liquid limit; I_p = plasticity index, G_s = specific gravity

Figure 4 shows the measured SWCCs (shown in symbols), the fitted SWCCs using Fredlund and Xing (1994) equation (shown in broken lines) and the predicted SWCC using Houston et al. (2006) method (shown in continuous lines) for the three subgrade soils. It can be seen that the SWCCs of the three soils are reasonably predicted and the predictions are close to the best-fit SWCCs using the Fredlund and Xing (1994) equation (Eq. 3).

Figure 5 shows the variation of the measured M_R with gravimetric water content for the three soils measured at the cyclic stress (σ_d) of 48.2 kPa and the confining stress (σ_c) of 27.6 kPa using different symbols. Figure 5 is shown as a relationship between the (M_R/M_{Rsat}) versus $(w-w_{opt})/(w_{sat}-w_{opt})$ relationships along with the soils' plasticity information (i.e. %clay and plasticity index I_p). The predicted M_R–w relationships using Eq. (3) and the predicted SWCCs in Fig. 4 are shown in solid lines. Figure 6 shows the measured (in symbols) and predicted (in broken lines) variation of the M_R with w.

The measured M_R–w relationships are nonlinear and sensitive to the clay content and I_p. The sensitivity of the M_R to the w increases with the clay content and the I_p. The M_R increases 50% from w_{sat} to w_{opt} (i.e., M_{Ropt} = 1.5 M_{Rsat}) for the low plastic TSC. For a similar change in water content, there is approximately 450% increase in M_R for the high plastic OLC (i.e. M_{Ropt} = 5.5 M_{Rsat}). Similar trends in observations are also reported in the literature for other fine-grained soils (e.g., Drumm et al. 1997; Khoury and Zaman 2004).

The lines described using Eq. (3) closely predict the measured nonlinear M_R–w relationships for all the soils. It should be noted that the predicted nonlinear M_R–w relationships are bounded by the measurements of M_{Rsat} and M_{Ropt} defined by Eq. (3) using only the SWCC predicted using Houston et al. (2006) method and

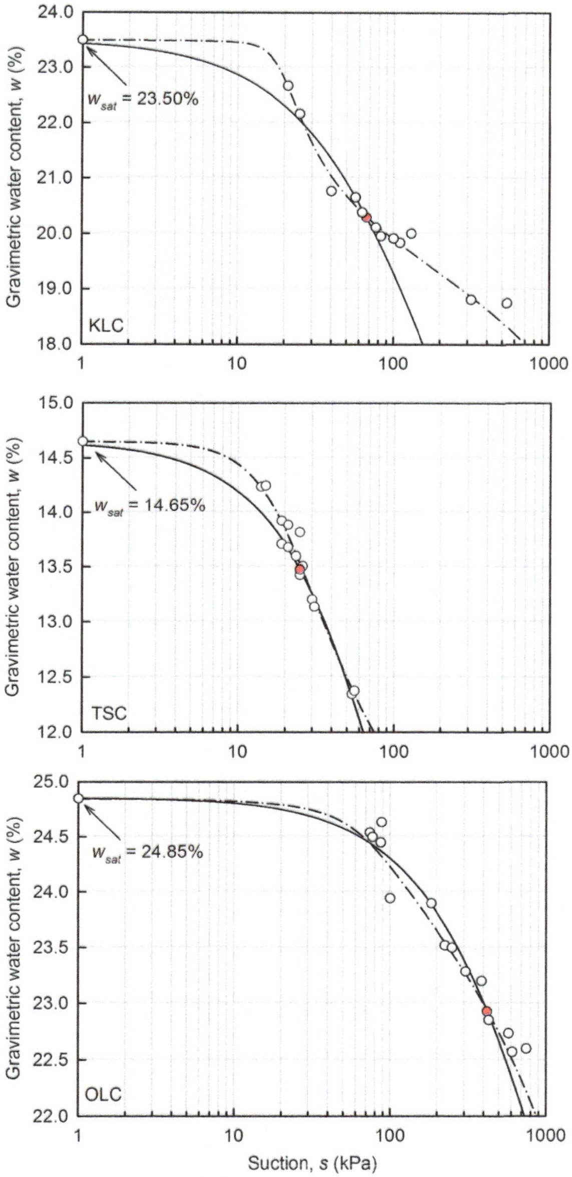

Fig. 4 Measured, fitted and predicted SWCCs for the three soils

the (s_{opt}, w_{opt}) measurements and a fitting parameter $\xi = 2.0$. If the M_{Rsat} and M_{Ropt} is described by the stress-dependent model such as Eq. (5) which is currently used in the MEPDG (ARA, Inc., ERES Consultants Division 2004) for predicting the stress-dependent M_R, the M_R–w–stress level relationships can also be predicted.

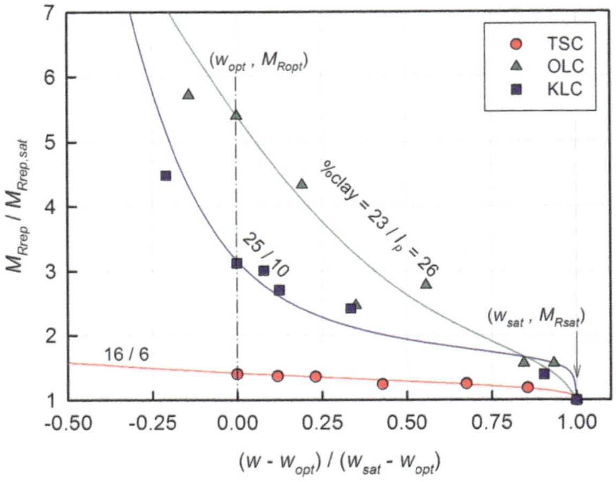

Fig. 5 Measured and predicted M_R for the three soils

$$M_R = k_1 p_a \left(\frac{\theta}{p_a}\right)^{k_2} \left(\frac{\tau_{oct}}{p_a} + 1\right)^{k_3}, \tag{5}$$

where θ = bulk stress, τ_{oct} = octahedral shear stress, p_a = atmospheric pressure, k_1, k_2 and k_3 = model parameters.

Figure 7 shows an example, for M_R–σ_d–w relationships of the three soils at σ_c = 27.6 kPa, the predictions provided by Eq. (2) using predicted SWCCs from Houston et al. (2006) method. The predicted surfaces are smooth and vary with the σ_d and w. They closely simulate the coupling between the M_R, external stress level and moisture content (i.e., M_R–external stress level relationship is influenced by the moisture content and M_R–moisture content external relationship is also influenced by the stress level) which is widely reported in the recent literature (Ng et al. 2013; Azam et al. 2013).

5 Summary and Conclusions

The influence of the environmental factors should be taken into account in the reliable design of pavements. Traditional pavement design methods are widely used in underdeveloped and developing countries and in some scenarios are also used in developed countries. These methods are based on shear strength parameters that do not properly address the failure mechanism of the pavements. Recently developed methods that are capable of taking account of the influence of environmental factors require the information of M_R–moisture content or M_R–suction relationships at various moisture content and suction levels are cumbersome, time consuming and expensive to obtain.

Fig. 6 Measured and predicted M_R–w relationships for the three soils

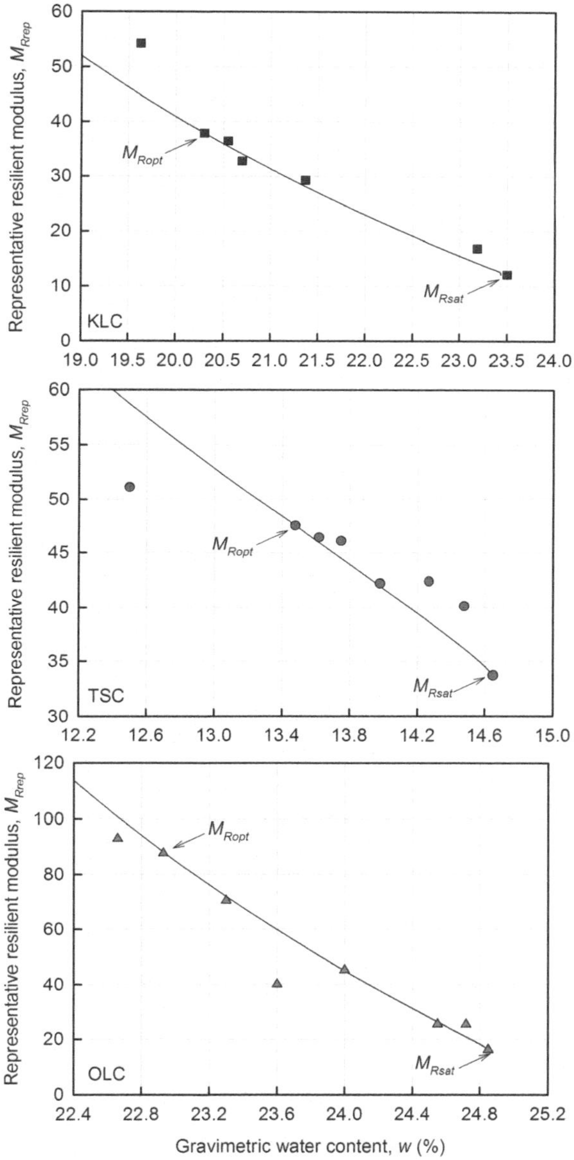

This paper introduces simplified and integrated approach that can be used to predict both the SWCC and the variation of the M_R with moisture content and suction for fine-grained subgrade soils. The SWCC and M_R are two pieces of information required for the rational design of pavement structures. This approach requires only the measurements of water and M_R at saturated condition (i.e., w_{sat} and M_{Rsat}) and optimum

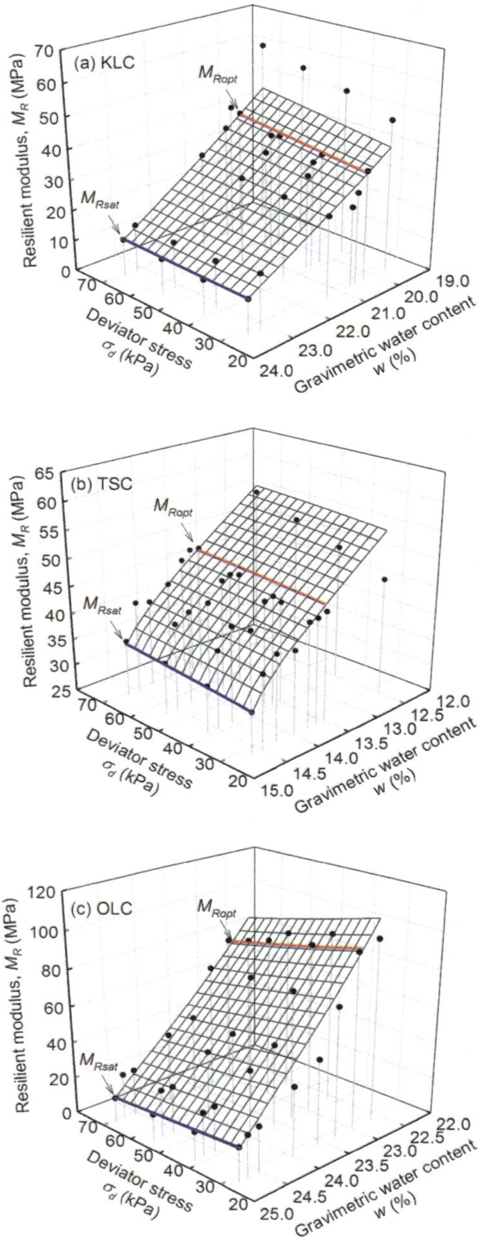

Fig. 7 Measured and predicted M_R–σ_d–w relationships of the three soils

moisture content condition (i.e., M_{Ropt}, s_{opt} and w_{opt}) for prediction. The proposed approach has been successfully validated for three subgrade soils from Canada. The integrated approach is simple, economical yet rational and hence can be used universally for addressing the present-day geotechnics challenges associated with the rational design of pavements.

Acknowledgements Authors gratefully acknowledge the funding received from the Natural Sciences and Engineering Research Council of Canada (NSERC) and the Ministry of Transportation of Ontario (MTO) which supported this study. Opinions presented in this study are essentially of the authors and may not necessarily reflect the views and policies of the NSERC and MTO.

References

ARA, Inc., ERES Consultants Division. 2004. *Guide for mechanistic-empirical design of new and rehabilitated pavement structures. Final report, NCHRP project 1-37A*. Washington, D.C.: Transportation Research Board.

Azam, A.M., D.A. Cameron, and M.M. Rahman. 2013. Model for prediction of resilient modulus incorporating matric suction for recycled unbound granular materials. *Canadian Geotechnical Journal* 50 (11): 1143–1158.

Bigl, S.R., and R.L. Berg. 1996. *Material testing and initial pavement design modeling: Minnesota road research project. CRREL report 96-14*. Hanover, N.H., USA: USACE Cold Regions Research and Engineering Laboratory.

Bosscher, P.J., and D.L. Nelson. 1987. Resonant column testing of frozen Ottawa sand. *Geotechnical Testing Journal* 10 (3): 123–134.

Cary, C.E., and C.E. Zapata. 2011. Resilient modulus for unsaturated unbound materials. *Road Materials and Pavement Design* 12 (3): 615–638.

Catana, M.C., S.K. Vanapalli, and V.K. Garga. 2006. The water retention characteristics of compacted clays. In *Proceedings of the unsaturated soils 2006*, 1348–1359.

Chin, K.B., E.C. Leong, and H. Rahardjo. 2010. A simplified method to estimate the soil-water characteristic curve. *Canadian Geotechnical Journal* 47 (12): 1382–1400.

Coronado, O., B. Caicedo, S. Taibi, A.G. Correia, H. Souli, and J.M. Fleureau. 2016. Effect of water content on the resilient behavior of non standard unbound granular materials. *Transportation Geotechnics* 7: 29–39.

Drumm, E.C., J.S. Reeves, M.R. Madgett, and W.D. Trolinger. 1997. Subgrade resilient modulus correction for saturation effects. *Journal of Geotechnical and Geoenvironmental Engineering* 123 (7): 663–670.

Fredlund, D.G. 2006. Unsaturated soil mechanics in engineering practice. *Journal of Geotechnical and Geoenvironmental Engineering* 132 (3): 286–321.

Fredlund, D.G., and A. Xing. 1994. Equations for the soil-water characteristic curve. *Canadian Geotechnical Journal* 31 (4): 521–532.

Han, Z., and S.K. Vanapalli. 2015. Model for predicting the resilient modulus of unsaturated subgrade soil using the soil-water characteristic curve. *Canadian Geotechnical Journal* 52 (10): 1605–1619.

Han, Z., and S.K. Vanapalli. 2016a. State-of-the-art: Prediction of resilient modulus of unsaturated subgrade soils. *International Journal of Geomechanics*. https://doi.org/10.1061/(ASCE)GM.1943-5622.0000631.

Han, Z., and S.K. Vanapalli. 2016b. Relationship between resilient modulus and suction for compacted subgrade soils. *Engineering Geology* 211: 85–97.

Houston, W.N., H.B. Dye, C.E. Zapata, Y.Y. Perera, and A. Harraz. 2006. Determination of SWCC using one point suction measurement and standard curves. In *Proceedings of the unsaturated soils 2006*, 1482–1493.

Huang, Y.H. 2004. *Pavement analysis and design*, 2nd ed. New Jersey: Prentice Hall.

Jong, D.T., P. Bosscher, and C. Benson. 1998. Field assessment of changes in pavement moduli caused by freezing and thawing. *Transportation Research Record: Journal of the Transportation Research Board* 1615: 41–48.

Khoury, N.N., and M.M. Zaman. 2004. Correlation between resilient modulus, moisture variation, and soil suction for subgrade soils. *Transportation research record, 1874, Transportation Research Board*, Washington, D.C., 99–107.

Khoury, C.N., R. Brooks, S.Y. Boeni, and D. Yada. 2013. Variation of resilient modulus, strength, and modulus of elasticity of stabilized soils with postcompaction moisture contents. *Journal of Materials in Civil Engineering* 25 (2): 160–166.

Lee, W., N.C. Bohra, A.G. Altschaeffl, and T.D. White. 1995. Resilient modulus of cohesive soils and the effect of freeze-thaw. *Canadian Geotechnical Journal* 32 (4): 559–568.

LeKarp, F., U. Isacsson, and A. Dawson. 2000. State of the art. I: Resilient response of unbound aggregates. *Journal of Transportation Engineering* 126 (1): 66–75.

Li, Q., X. Ling, and D. Sheng. 2015. Elasto-plastic behaviour of frozen soil subjected to long-term low-level repeated loading, part I: Experimental investigation. *Cold Regions Science and Technology* 125: 138–151.

Ng, C.W.W., and C. Zhou. 2014. Cyclic behaviour of an unsaturated silt at various suctions and temperatures. *Géotechnique* 64 (9): 709–720.

Ng, C.W.W., C. Zhou, Q. Yuan, and J. Xu. 2013. Resilient modulus of unsaturated subgrade soil: Experimental and theoretical investigations. *Canadian Geotechnical Journal* 50 (2): 223–232.

Perera, Y.Y., C.E. Zapata, W.N. Houston, and S.L. Houston. 2005. Prediction of the soil–water characteristic curve based on grain-size-distribution and index properties. *Geotechnical Special Publication* 130: 49–60.

Salour, F., S. Erlingsson, and C.E. Zapata. 2014. Modelling resilient modulus seasonal variation of silty sand subgrade soils with matric suction control. *Canadian Geotechnical Journal* 51 (12): 1413–1422.

Sawangsuriya, A., T.B. Edil, and C.H. Benson. 2009. Effect of suction on resilient modulus of compacted fine-grained subgrade soils. *Transportation research record, 2101, Transportation Research Board*, Washington, D.C., 82–87.

Zapata, C.E., D. Andrei, M.W. Witczak, and W.N. Houston. 2007. Incorporation of environmental effects in pavement design. *Road Materials and Pavement Design* 8 (4): 667–693.

Characterization of Sensitive Soft Clays for Design Purposes

V. Thakur

1 Introduction

Decision-making and design processes in geotechnical engineering rely heavily on site characterization. The best approach for the site characterization of fine-grained soils (silts and clays) involves a combination of in situ testing and laboratory testing of undisturbed samples. Site characterization is often performed in two steps. The first step includes determinations of the soil profile and hydrological conditions, which involve the identification of the soil types and their in situ state, i.e., natural water content, unit weight, consistency, void ratio, etc., together with the location of the ground water table and pore pressure profile along the depth. The second step entails an estimation of the relevant engineering properties that quantify the properties of the soils and are needed for design.

Marine clays that have been subjected to leaching by a flow of fresh water, whereby the salt ion concentration in the pore water is reduced to 3–5 g/L, are called leached marine clays. Such clays are characterized using its sensitivity (S_t), which is the ratio between undrained shear strength (c_u) and remolded shear strength (c_{ur}). The engineering characterization of leached marine clays is a challenging task because this type of material remolds upon the slightest disturbance. Because of this, undrained shear strength, stiffness, and pre-consolidation stress are often underestimated.

Hence, high quality sampling is essential to get a proper understanding of the engineering behavior of such clays; one alternative is to do block sampling (see Fig. 1). In short, characterization based on high-quality sampling and in situ testing contributes to the safer and economical constructions in or on a sensitive soft clay

V. Thakur (✉)
Geotechnical Engineering, Norwegian University of Science and Technology,
Trondheim, Norway
e-mail: vikas.thakur@ntnu.no

K. Ilamparuthi and R. G. Robinson (eds.), *Geotechnical Design and Practice*,
Developments in Geotechnical Engineering,
https://doi.org/10.1007/978-981-13-0505-4_9

deposit. Accordingly, this paper concentrates on the strength characterization of sensitive soft clays using the advanced laboratory and field-testing methods.

2 Laboratory Testing

The most reliable laboratory method to assess undrained shear strength (compression or extension) is triaxial testing. Oedometer test(s) should be performed before triaxial testing so that pre-consolidation pressure of the material is known. Pre-consolidation pressure can be measured by different methods, including Casagrande's method, Janbu's method, or Salfor's method. A typical triaxial and odometer result for a soft clay is shown in Fig. 2. Anisotropically consolidated triaxial undrained compression tests (CAUC) and oedometer tests were performed on 54 and 73 mm diameter tube samples of the Rissa clay, and the results are compared with those obtained from block samples. For the tube samples, it was found that the 73 mm tube sampler gives a better result than the 54 mm. The reason for the is because of overstraining that occurs when the sampler penetrates the soil and when the sample is extracted from the tube prior to testing. Additionally, the friction between the clay and the steel tube causes the outer zone of the sample to become remolded. Block sampling avoids this type of disturbance; however, stress

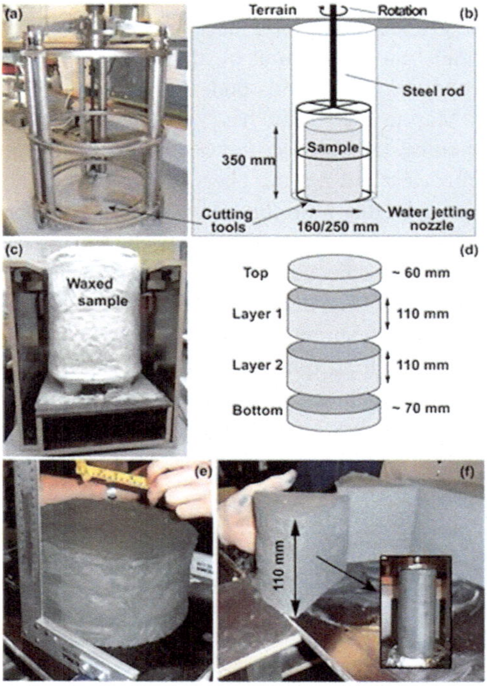

Fig. 1 **a** Sherbrooke block sampler at NTNU, **b** schematic view of a block sample being card, **c** waxed sample, **d** schematic view of a sliced sample, **e** a block sample slice and **f** a piece of clay from block sample (*Photo* H. Amundsen, NTNU)

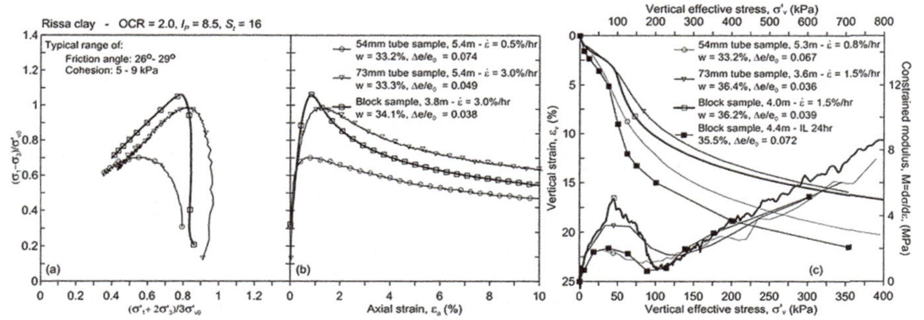

Fig. 2 (left) Undrained triaxial test results (right) incremental and constant rate of strain odometer test results for a sensitive soft clay. The specimen were sampled using tube samplers (54 and 73 mm diameter) and the Sherbrooke block sampler. Here OCR is the over consolidation ratio, *Ip* is the plasticity index, S_t is the sensitivity. *w* is the water content, $\Delta e/e_o$ is the change in the void ratio, and the strain rate is defined by $\dot{\varepsilon}$ (Amundsen et al. 2016)

Fig. 3 Stress relief and sample quality of block samples on sensitive soft clays

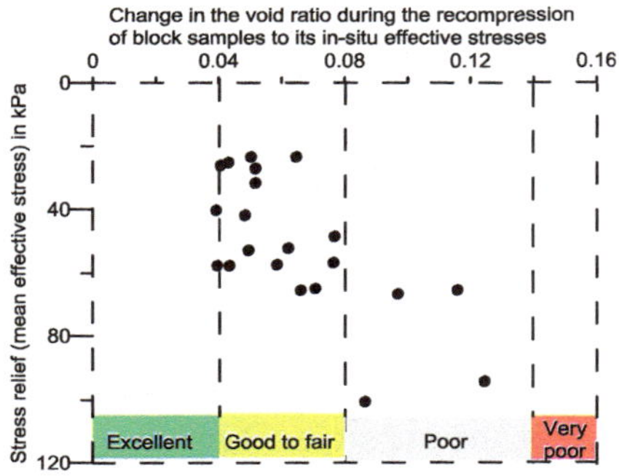

relief may become an issue if the sample is extracted from a great depth. With increasing stress relief, sensitive soft clay samples usually becomes poorer in quality. As illustrated in Fig. 3, even block samples may yield poorer quality if they are subjected to stress relief. A low plastic soft clay losses most of its residual suction (negative pore pressure induced due to the stress relief) and swells inside the cylinder before testing. This fact demands an understanding of how quickly the suction may be lost. The answer to this issue lies in the sizes and the inter particle bonding between clay particles. Permeable materials have a lower tendency to exhibit suction and higher tendency to lose it during and after testing. Recent work by Amundsen et al. (2016) suggests that sensitive soft clays may lose its suction in a matter of some minutes to hours depending on the constitution of the material. Low plastic sensitive soft clays are rich in silt and they tend to lose their suction quickly and cause swelling of the sample. Despite careful reconsolidation, such material will always result in poorer sample quality.

In Fig. 2 an incremental loading (IL with 24 h load steps) oedometer test and a constant rate of strain (CRS—1.5%/h) odometer test from the same block sample are also compared. The pre-consolidation stress is 95 kPa for the CRS test and 80 kPa for the IL test, which corresponds to a 16% decrease. The pre-consolidation stress from CRS oedometer tests is rate dependent and may result in a high pre-consolidation stress at high strain rates. On the other hand, the 24 h load duration in IL tests causes creep deformations in the soil, which may yield a lower pre-consolidation pressure. This will impact negatively if one estimates undrained shear strength using the SHANSEP formula.

Knowledge about the in situ effective stress (pore pressure measurement) at the sample collection depths for the triaxial tests is valuable. In this way, one can estimate an accurate over consolidation ratio (OCR), which can also be used to find K_0' for use in triaxial testing. Estimation of the correct K_0' is demanding. However, one can use the approach suggested by Brook and Ireland in 1965 that provides a relatively easy way to estimate K_0' based on soil plasticity (I_p) and the over consolidation ratio (OCR). Caution is needed in using a relatively high K_0' because this will result in a higher average effective stress in the sample, resulting in a high active undrained shear strength (c_{uC}). As in the odometer test, the results of a triaxial test depend on the strain rate (rate dependence). An increase in the strain rate generally results in increased maximum undrained strength and brittle behavior. The literature suggests that there might be a factor of approximately 1.5 in the estimate c_{uC} for very fast rate to a very slow rate. Commonly used strain rates in Norway have been 0.7–3.0%/h. In practice, there is little difference between these experiments, and normally we do not distinguish between these strain rate and strain softening behavior.

While estimating c_{uC} from triaxial tests, a distinction must be made between dilating and contracting behavior. For the tests exhibiting contracting behavior (positive excess pore pressure build-up during testing), it is recommended to obtain c_{uC} at the maximum measured shear resistance. In many ways, the stress path summarizes all major considerations regarding sampling and sample disturbance. A test that first exhibits a contracting behavior but then shows dilation is an indication of sample disturbance. In such cases, it is advisable to not go beyond the c_{uC} that shows the point representing the transition between contractancy and dilatancy. Indeed, dilating materials (highly over-consolidated clays) often attain high strengths; however, this is normally related to high strains. In these cases, it is recommended that the c_{uC} is defined by a given strain, for example 10%.

3 In Situ Characterisation

The methods applied in field characterization of soft sensitive clays materials must be chosen based on a cost–benefit perspective, the applicability of the methods for the actual ground conditions and the general use of soil data in the project. For use in current practice, it is important to present recommendations based on the experiences and observations made with various detection methods. In particular, this is valid for the resistivity methods CPTU, field vane shear R-CPTU, ERT, and AEM, where limited experience exists from practical use.

The most common field test to estimate c_{uC} is the CPT apart from vane shear testing. In Scandinavian countries, the CPT with pore pressure measurements (CPTU) is considered to be the common method for estimating active undrained shear strength (c_{uC}) due to its ability to provide information that can help in establishing a continuous c_{uC} measurement along with the depth. The value of c_{uC} is based on three cone factors: N_{kt}, $N_{\Delta u}$, and N_{ke}.

The total tip resistance based c_{uC} is calculated as,

$$c_{uC} = \frac{q_T - \sigma_{vo}}{N_{kt}} \tag{1}$$

Here, q_T is tip resistance and s_{vo} is the total vertical pressure. The pore pressure measurement based c_{uC} is calculated as,

$$c_{uC} = \frac{u_2 - u_o}{N_{\Delta u}} \tag{2}$$

Here u_2 is the measured pore pressure and u_o is the in situ pore pressure. The effective tip resistance based c_{uC} is calculated as,

$$c_{uC} = \frac{q_T - u_2}{N_{ke}} \tag{3}$$

More information regarding the testing and interpretation of the CPTU can be obtained in the literature. It is clear from Eqs. (1)–(3) that c_{uC} will depend on the cone factors. Several correlations exist for the N_{kt}, N_{ke}, and $N_{\Delta u}$ parameters to calculate c_{uC}. The widely accepted method in Norway is compared the CPTU with c_{uC} from block samples and suggested the cone factors N_{kt} and $N_{\Delta u}$ based on the OCR and soil plasticity index (I_p). The N_{ke} parameter was correlated with B_q (pore pressure parameter). As listed below, suggestions were made for soil having sensitivity less than 15

$$N_{kt} = 7.8 + 2.5 \cdot \log \text{OCR} + 0.082 \cdot I_p \tag{4}$$

$$N_{\Delta u} = 6.9 – 4.0 \cdot \log \text{OCR} + 0.07 \cdot I_p \tag{5}$$

$$N_{ke} = 11.5 – 9.05 \cdot B_q \tag{6}$$

$$B_q = 0.88 – 0.51 \cdot \log \text{OCR} \tag{7}$$

Similar suggestions have been made for sensitive soft clays with sensitivity greater than 15. Before selecting a value for the cone factors, the basis of selection and the available information in terms of routine investigations, in situ pore pressure, odometer, and triaxial tests are needed, and the choice of the interpretation method needs consideration and justification. Moreover, the validity of the interpretation of factors N_{kt} and $N_{\Delta u}$ and especially for $N_{\Delta u}$ at low B_q must be considered (Thakur 2016).

Vane shear testing is another approach to obtained DSS strength at a desired depth. The vane shear test has been widely used as an in situ test device in Norway from the sixties to the eighties. However, the last decades, its popularity has decreased, partly because of the increasing popularity of the CPTU-test, but also because of uncertainties related to interpretation of the vane shear test. It is possible to establish a strong relation between the active undrained shear strength and the undrained shear strength as interpreted from the vane shear test as a function of the plasticity index. If is further seen that there is a potential for deducing OCR from the vane test. There is not a one-to-one relation between sensitivity as measured from the vane test and by the falling cone test in the laboratory. The vane appears to measure too high values for the remolded undrained shear strength. Work on the relation between OCR and the VST exists in the literature. One of these, Mayne and Mitchell (1988), use the SHANSEP framework which is based on the relation between undrained shear strength and OCR: $(c_u/\sigma_{v0}') = a\ \mathrm{OCR}^m$. By knowing the undrained shear strength from the VST together with appropriate factors (a and m), it is possible to determine OCR. A considerable dataset is gathered in Mayne and Mitchell (1988) to investigate the validity of this approach. The dataset of Mayne and Mitchell (1988) show a clear trend of increasing c_u/σ_{v0}' for increasing OCR. Lines for four α-values ($1/a$) are included in the figure. The three Norwegian clays tested here lies within the data trend. α-values in the range of 6–8 are indicated. $m = 1$ is assumed in this interpretation (Gylland et al. 2016).

In recent years, electric field vane has been used in Norway. One key aspect of the electric field vane is to establish mobilized direct shear resistance (c_{uD}) curve with vane rotation. The mobilized shear resistance is interpreted using $c_{uD} = 6T/(7\pi D^3)$ where T is the torque. D is the vane diameter. The usefulness of such plot is in term of estimation of remolding energy (Thakur and Degago 2013), studies related to shear band and progressive failure (Gylland et al. 2016). A typical interpreted result from an electric field vane test is shown in Fig. 4.

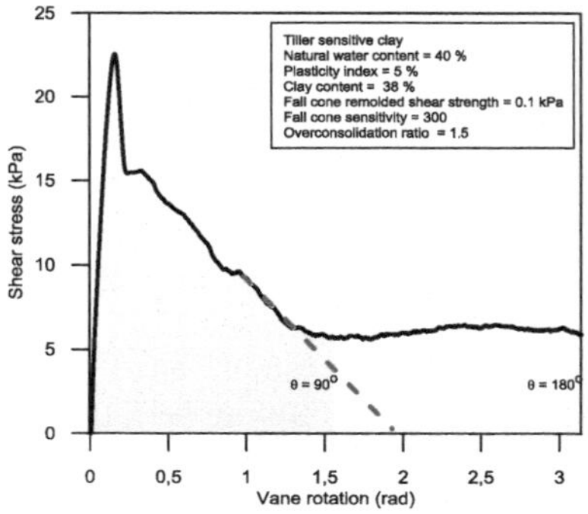

Fig. 4 Mobilized shear stress versus vane rotation for a sensitive soft clay. Here, θ refers to vane rotation. The dotted line is the anticipated shear stress-vane rotation curve when $\theta > 90°$

4 Design Considerations

In Norway, total stress analysis ($\phi = 0$) is mostly used. The validity of the $\phi = 0$ analysis for calculating the "end of construction" stability of slopes, cuttings, the bearing capacity of footings, and fillings on clay has been very well adopted in Norway and elsewhere. A pragmatic approach to estimate the factor of safety from total stress based on the over consolidation ratio (OCR) because it seems to be a major factor influencing the factor of safety in total stress analyses.

$$\text{Factor of Safety} \sim 0.9\ \text{OCR}$$

Assessing the pore pressure situation is the biggest challenge using the effective stress analysis. The effect of natural variations (seasonal variations, extreme precipitation) must be accounted for in the form that is considered the most unfavorable condition based on control over time with an extrapolation of efficacy on the pore pressure situation related to extreme precipitation. Climate-induced changes must be taken into consideration in relation to changes in future ground water levels. The basic premise of adding effective stress analysis as a basis for the stability assessment of a natural slope is that it is not exposed to the geological or artificial activities resulting in a deterioration of stability.

Strength anisotropy must be considered when short-term stability calculations are performed using the total stress based approach ($\phi = 0$). A recommendation practice for use of strength anisotropy is proposed by Thakur et al. (2017) and is shown in Fig. 5. Thakur et al. (2017) further show that the variation of the strength anisotropy factors in the range ($c_{uD}/c_{uC} = 0.65$, $c_{uE}/c_{uC} = 0.33$) and ($c_{uD}/c_{uC} = 0.60$, $c_{uE}/c_{uC} = 0.30$) leads to a 4–8% change in the calculated factor of safety. The size of the change is a direct result of the relative distribution between the active, direct and passive zone for the sliding surfaces. For the stability analysis of slopes, the passive zone normally constitutes a small percentage of the sliding surface.

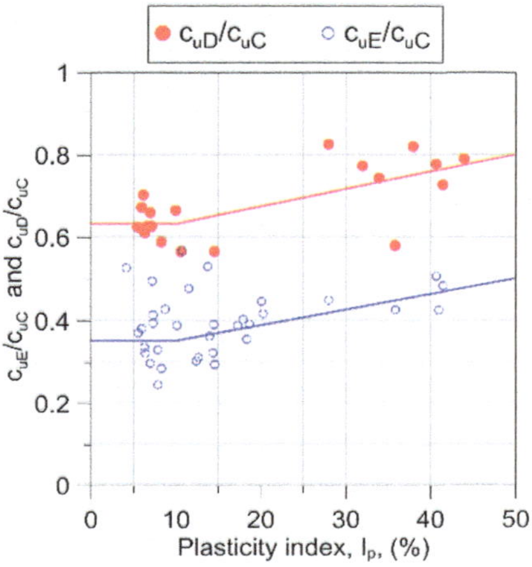

Fig. 5 Recommendation for the use of DSS strength anisotropy (c_{uD}/c_{uC}) and Passive strength anisotropy (c_{uE}/c_{uC}). Here C_u refers to the undrained shear strength and subscripts C, E, D are the compression, the extension and the direct shear condition (Thakur et al. 2017)

5 Closing Remarks

In this paper, an effort has been made to illustrate the possibilities and challenges that one may encounter while doing the characterization of sensitive soft clays in situ or in the laboratory. Reliability of the derived engineering parameters of sensitive soft clays increases with the quality of samples and in situ testing methods.

Acknowledgements Author would like to acknowledge Dr. Ander Gylland from Multiconsult AS and Mrs. Helene Amundsen, Ph.D. student at the NTNU Norway, for providing resources.

References

Amundsen, H.A., V. Thakur, and A. Emdal. 2016. Sampling induced disturbance in the block samples of low plastic sensitive soft clays. In *17th NGM*, Iceland.

Gylland, A., R. Sandven, A. Emdal, and V. Thakur. 2016. Extended interpretation basis for the vane shear test. In *17th NGM*, Iceland.

Mayne, P., and J. Mitchell. 1988. Profiling overconsolidation ratio in clays by field vane. *Canadian Geotechnical Journal* 25: 150–157.

Thakur, V. 2016. Characterisation of sensitive soft clays for design purposes. In *IGC-2016*, 15–17 Dec 2016, IIT Madras, India.

Thakur et al. 2017. *Recommended practice for use of strength anisotropy*. 2nd IWLSC, Springer book series on natural hazards.

Thakur, V., and S.A. Degago. 2013. Disintegration energy of sensitive clays. *Géotechnique Letters* 3 (1): 21–25.

Dynamic SSI of Monopile-Supported Offshore Wind Turbines

S. Bhattacharya, G. Nikitas and N. Vimalan

1 Introduction

1.1 Loads on Offshore Wind Turbine Foundations

Designing foundations for OWTs are challenging as these are dynamically sensitive structures in the sense that natural frequencies of these structures are very close to the forcing frequencies, see for example Bhattacharya (2014) and Bhattacharya and Adhikari (2011). A designer apart from predicting the global natural frequency of the structure, must also ensure that the overall natural frequency due to dynamic soil–structure interaction does not shift towards the forcing frequencies. Further details can be found in references Lombardi et al. (2013) and Bhattacharya et al. (2013).

Figure 1 shows a schematic diagram of a monopile-supported wind turbines with the main loads acting on them. The figure also shows the characteristics of the mudline bending moment acting on the pile head. Typically, in shallow to medium deep waters, the wind thrust loading at the hub will produce the highest cyclic overturning moment at the mudline. However, the frequency of this loading is extremely low and is in the order of magnitude of 100s (see Fig. 1). Typical period of wind turbine structures being in the range of about 3 s, see Arany et al. (2016), no resonance of structure due to wind turbulence is expected resulting in cyclic soil–structure interaction. On the other hand, the wave loading will also apply overturning moment at the mudline and the magnitude depends on water depth, significant wave height and peak wave period. Typical wave period will be in the

S. Bhattacharya (✉) · G. Nikitas
Surrey Advanced Geotechnical Engineering (SAGE) Laboratory,
University of Surrey, Guildford, UK
e-mail: S.Bhattacharya@surrey.ac.uk

N. Vimalan
VJTech, Reading, UK

K. Ilamparuthi and R. G. Robinson (eds.), *Geotechnical Design and Practice*,
Developments in Geotechnical Engineering,
https://doi.org/10.1007/978-981-13-0505-4_10

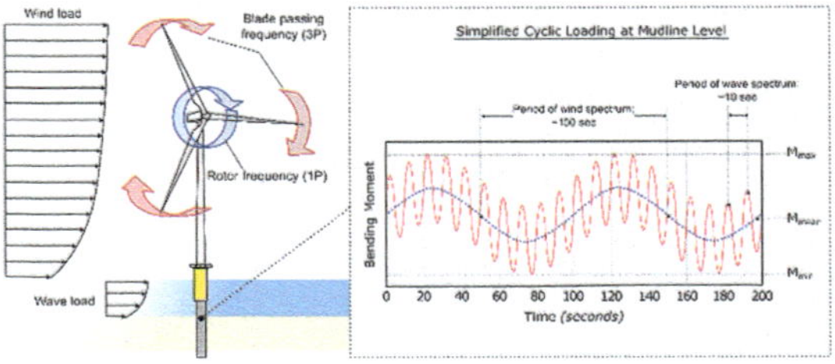

Fig. 1 Loads acting on a typical offshore wind turbine foundation and typical mudline moment

order of 10 s (for North Sea) and will therefore have dynamic soil–structure interaction.

A calculation procedure is developed in Arany et al. (2016) and the output of such a calculation will be relative wind and the wave loads and an example is shown in Fig. 1. It is assumed in the analysis that the wind and wave are perfectly aligned which is a fair assumption for deeper water further offshore projects (i.e. fetch distance is high). Analysis carried out by Arany et al. (2014) showed that the loads from 1P and 3P are orders of magnitude lower than wind and wave but they will have highest dynamic amplifications. The effect of dynamic amplifications due to 1P and 3P will be small amplitude vibrations. Resonance has been reported in operational wind farms in German North Sea, see Hu et al. (2014). Furthermore, there are added soil–structure interactions due to many cycles of loading and the wind–wave misalignments. Typical estimates will suggest that offshore wind turbine foundations are subjected to 10–100 million load cycles of varying amplitudes over their lifetime (25–30 years). The load cycle amplitudes will be random/irregular and have broadband frequencies ranging several orders of magnitudes from about 0.001–1 Hz.

1.2 Dynamic and Cyclic Soil–Structure Interaction in Offshore Wind Turbines

Based on the discussion above, the soil–structure interaction can be simplified into two superimposed cases, see Nikitas et al. (2016) and is discussed below:

(a) Cyclic overturning moments (typical frequency of 0.01 Hz) due to lateral loads of the wind acting at the hub. This will be similar to a "*fatigue type*" problem for the soil and may lead to strain accumulation in the soil giving rise to progressive tilting. Due to wind and wave load misalignment, the problem can be bi-axial. For example, under operating condition, for deeper water and further offshore sites, wind–wave misalignment will be limited for most practical scenarios. Wave loading, on the other hand, will be moderately dynamic as the frequency of these loads is close to natural frequency of the whole wind turbine system (typical wind turbine frequency is about 0.3 Hz).

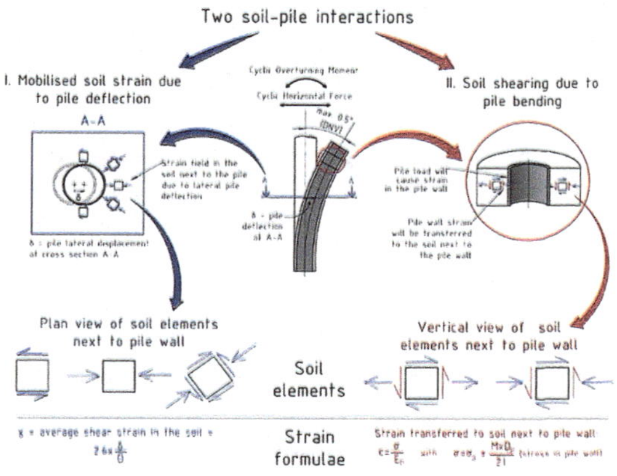

Fig. 2 Two types of soil-pile interaction on a monopile-supported wind turbine

(b) Due to the proximity of the frequencies of 1P, 3P, wind, and wave loading to the natural frequency of the structure, resonance in the wind turbine system is expected and has been reported in German Wind farm projects Hu et al. (2014). This resonant dynamic bending moment will cause strain in the pile wall in the fore-aft direction which will be eventually be transferred to the soil next to it. This resonant type mechanism may lead to compaction of the soil in front and behind the pile (in the fore-aft direction).

Deformation of the pile under the action of the loading described in Fig. 1 will lead to three-dimensional soil-pile interaction as shown schematically in Fig. 2. Simplistically, there would be two main interactions: (a) due to pile bending (which is cyclic in nature) and the bending strain in the pile will transfer (through contact friction) strain in the soil which will be cyclic in nature; (b) due to lateral deflection of the pile there will be strain developed in the soil around the pile. Figure 2 shows a simple methodology to estimate the levels of strains in a soil for the two types of interactions and is given by Eqs. (1) and (2). The average strain in the soil at any section in a pile due to deflection can be estimated using as follows:

$$\gamma = 2.6\frac{\delta}{D_{\mathrm{P}}}, \tag{1}$$

where δ is the pile deflection at that section (for example A-A in Fig. 2) and D_{P} is the pile diameter.

On the other hand, the shear strain in the soil next to the pile due to pile bending, can be estimated using Eq. (2).

$$\gamma_1 = \frac{M \times D_{\mathrm{P}}}{2 \times I \times E_{\mathrm{P}}}, \tag{2}$$

where M is the bending moment in the pile, I is the second moment of area of pile and E_P is the Young' Modulus of the pile material. It must be mentioned that Eq. (2) assumes that 100% of the strain is transmitted to the soil which is a conservative assumption and calls for further study. In practice, this will be limited to the friction between the pile and the soil.

1.3 Aim and Scope of the Paper

The aim of this paper is to study the cyclic soil–structure interaction through Cyclic Simple Shear apparatus where many testing parameters were changed. Tests have been carried out on a silica sand on three different relative densities (25, 50 and 75%) where 50,000 or more cycles of uniform cyclic strain of different amplitudes were applied and also under three different vertical stresses. The intention is to develop a framework of understanding which can be used to develop a methodology for prediction of long-term performance.

2 Cyclic Simple Shear Apparatus, Materials and Method of Testing

Cyclic simple shear apparatus, as shown in Fig. 3 is used for testing cylindrical samples of 50 mm in diameter and 20 mm in height. The apparatus is capable of applying vertical and horizontal loads using two electro-mechanical dynamic actuators. External LVDTs were also used to record displacements and to verify the effectiveness of feedback control. Loads up to ±5 kN can be applied in two directions with horizontal travel up to 25 mm and vertical travel of 15 mm. These are sufficient to study the effects of large strain levels applied to the soil. This therefore allows to study effects of cyclic shear stress under drained and undrained conditions. The loads can be applied at frequencies of up to 5 Hz. RedHill 110 Sand (poorly graded fine grained silica sand) was tested in this research as this soil

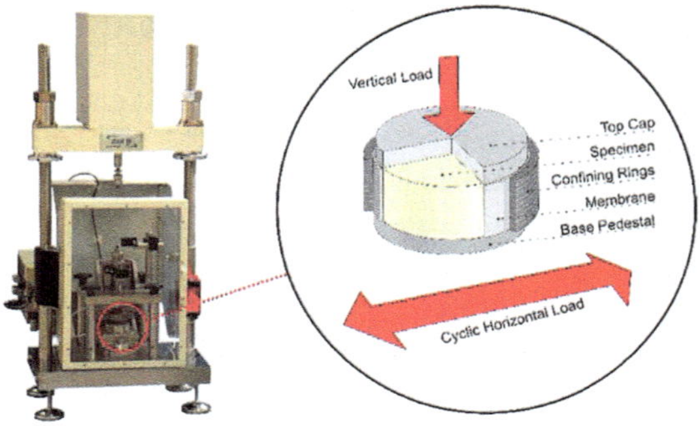

Fig. 3 Dynamic/cyclic simple shear apparatus with details of the sample

has been used to carry out scaled model tests on different types of foundations. The sand has a specific gravity, G_s of 2.65 and minimum and maximum void ratio of 0.608 and 1.035 respectively.

Strain controlled tests were carried out on a loose to medium dense sand (Dr = 50%) whereby the shear strain amplitudes were ranging from 0.02 to 10% and 50,000 cycles were applied. The following types of tests were carried out: (a) Cyclic shearing at seven different shear strain amplitudes (γ_c) with constant vertical stress of 100 kPa—Series A; (b) Cyclic shearing at three different vertical stresses (σ_v) with shear strain amplitude of 0.2%—Series B; (c) Cyclic shearing at two relative densities with shear strain amplitude of 0.2% and a constant vertical stress of 100 kPa—Series C.

3 Test Results and Discussion

3.1 Effect of Shear Strain Amplitude on Accumulated Strain and Shear Modulus

Figure 4a shows the average vertical strain accumulation with number of cycles for seven shear strain amplitude tests for a vertical consolidation stress of 100 kPa plotted in a log scale. The rate of vertical strain accumulation reduces with number of cycles. Also, the accumulated vertical strain increases with increasing shear strain amplitude. Cyclic Stress Ratio (CSR) i.e. (τ_{max}/σ_v') is increased in Series A tests. It may be observed that with increasing CSR, the rate of accumulation of

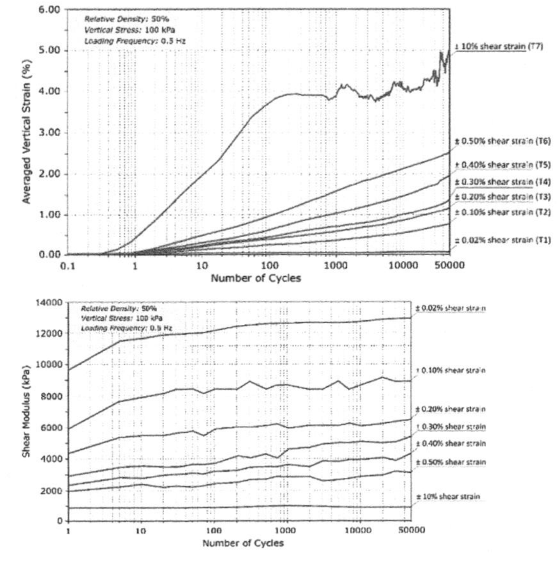

Fig. 4 **a** Vertical strain accumulation at vertical consolidation stress of 100 kPa; **b** Variation of shear modulus for different cyclic shear strain plotted against the number of cycles

vertical strain increases. This observation is similar to Silver and Seed (1971) where cyclic simple shear experiments were carried out on crystal silica sand for approximately 300 cycles. Figure 4b shows the shear modulus of the soil plotted for seven different shear strains. As expected, the initial shear modulus (i.e., before the cyclic stresses are applied) is dependent on the shear strain amplitude and reduces with increasing strain. The tests showed that the shear modulus generally increases with cycles of load.

In the context of Offshore Wind Turbine, the Soil–structure Interaction will differ at different depths and can be described by CSR (Cyclic Stress Ratio) which is essentially the ratio of shear stress to the vertical stress. Scaling laws deduced by Bhattacharya et al. (2011), Lombardi et al. (2013) showed CSR is proportional to the average shear strain around the pile. It is also quite clear that the soil at shallower depths is subjected to higher shear stress and low vertical stress giving a higher value of CSR. In this context, it must be mentioned that Abdel-Rahman et al. (2014) used cyclic simple shear test to predict cyclic capacity degradation of axially loaded piles which are applicable for small diameter pile supporting a jacket structure.

3.2 *Effect of Vertical Consolidation Stress*

The effect of vertical stress on the settlement of sand was studied by varying the vertical stress for constant shear strain amplitude of 0.2%. Figure 5a shows that the vertical strain accumulation decreases with higher vertical stress. Figure 5b on the other hand plots the change in shear modulus with cycles of loading for increasing vertical consolidation stress suggesting shear modulus increase with increasing depth which is consistent with the expectations and observations of Series A tests.

3.3 *Effect of Density*

The effect of density on behaviour of sand during cyclic loading was investigated for 3 relative densities (25, 50 and 75%) for a constant shear strain amplitude of 0.2% and constant vertical stress of 100 kPa. Vertical strain reduces with increasing relative density as shown in Fig. 6a. In other words, sands with lower relative density will have a higher strain accumulation. Similar observations were also reported by Silver and Seed (1971) but for much lower number of cycles. Shear modulus also increases with cycles of loading.

Fig. 5 **a** Vertical strain accumulation for different vertical consolidation stress applied at constant shear strain amplitude of 0.2%; **b** Shear modulus with number of cycles at vertical consolidation stress of 100 kPa

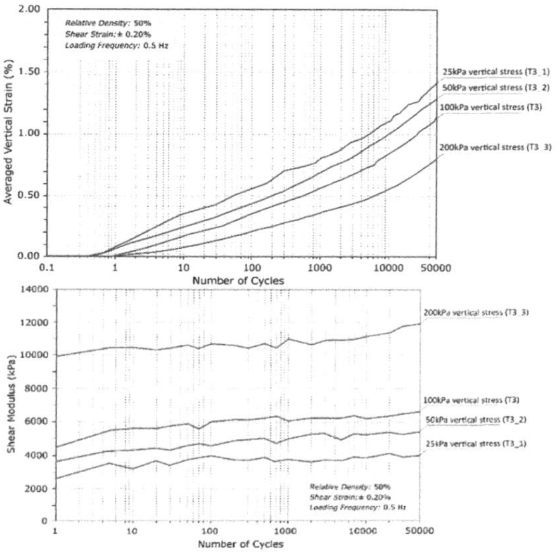

Fig. 6 **a** Vertical strain for varying relative densities at constant shear strain amplitude of 0.2% and vertical stress of 100 kPa; **b** Shear modulus for varying relative densities at constant shear strain amplitude of 0.2% and vertical stress of 100 kPa

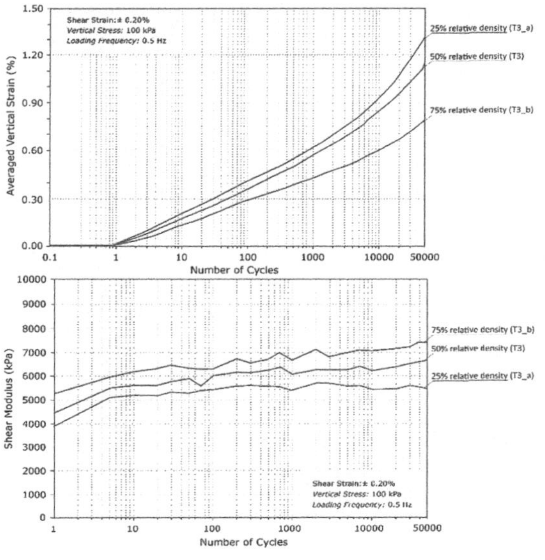

4 Discussion and Conclusions

The reported cyclic simple shear test under different stress conditions (different strain amplitudes, vertical stress or density) showed that shear modulus increases with cycles of loading under drained condition. The increase of shear modulus is pronounced in the first few hundred cycles and then it stabilises.

This result is consistent with the scaled model tests carried out on wind turbine foundations either in centrifuge [10] or 1-g and Discrete Element Method (DEM) study [10]. While offshore wind turbine structures are designed for an intended life of 25–30 years, little is known about their long-term dynamic behaviour under millions of cycles of loading. While monitoring of existing offshore wind turbine installations is a possibility and can be achieved at a reasonable cost, full scale testing is very expensive. An alternative method is to carry out a carefully planned scaled dynamic testing to understand the scaling/similitude relationships which can be later used for interpretation of the experimental data and also for scaling up the results to real prototypes. Leblanc et al. (2009) and later Cox et al. (2014) proposed an expression for change in foundation stiffness with number of cycles as shown by Eq. (3).

$$K_N = K_0 + A_\mathrm{k} \ln(N), \tag{3}$$

where K_0 is the initial foundation stiffness and N is the number of cycles, A_K is a constant depending on the problem (load directionality and magnitude). However, the change in monopile stiffness is closely (if not solely) linked to change in soil stiffness. Methods based on numerical analyses have been proposed by to analyse the soil–structure interaction issues. Based on a number of experimental investigations on monopiles and caissons, Eq. (4) has been proposed to predict the accumulation of rotation with number of cycles [10] based on the data collected and subsequently best fitted.

$$\frac{\Delta\theta(N)}{\theta_\mathrm{S}} = T.N^\alpha, \tag{4}$$

where N is the number of cycles and T is function of two parameters: ξ_b which specifies ratio of the maximum moment (M_max) to the static moment capacity, and ξ_c which represents the relative directionality (where $\xi_c = -1$ represents a symmetrical two-way loading and $\xi_c = 0$ represents a purely one-way regime). For detailed discussion about the function T, see [10] where the function is not presented in closed form but rather in figures.

For each loading condition the accumulation or retention of rotation could be assessed. The resulting change in rotation $\Delta\theta(N)$ could then be normalised with respect to the static rotation of the foundation [θ_s]. The fitting parameter α ranges between 0.18 and 0.39 and is obtained through scaled model tests where standard laboratory tests were used.

The next section highlights the main limitations of the above method: Equation (4) predicts a continuous increase of tilting with cycles of loading which seems to be physically unrealistic and may lead to overestimation of the tilt and uneconomic foundation design.

It is difficult, if not impossible, to reduce many millions of cycles of complex load patterns (wave, wind, 1P and 3P) into a single amplitude (T) and cycle count value (N) number. The method may work if it is calibrated which seems a formidable task.

The load acting on the foundation is a combination of different amplitudes arising from four different loads and may act in two different planes. However, the above method cannot take into account wind and wave misalignment. Apart from the above, wind turbines may vibrate in more than one direction and these directions will be different even in a single day.

The method is calibrated for laboratory sand and real soil will have a very different behaviour. Therefore using Eq. (4) for real problems needs further thoughts.

4.1 Possible Use of Element Tests for Long-Term Prediction

The element tests presented in this paper shed light on the cyclic soil–structure interaction and can be used to predict the long-term change in soil stiffness under different combination of loads. While there is continual increase in accumulated strain, there is a flattening of the shear modulus suggesting non-progressive (non-monotonous) tilting of the foundation. Considering Fig. 4a, strain level depicted in tests T_7 will have a very low probability of occurrence in the lifetime while strain levels in T_1 will have a high probability of occurrence in normal operating conditions. Again, based on the understanding developed on the particular soil from the site (say Fig. 4a), one can find the damage equivalence (for example accumulation of strain) of N_1 cycles of a particular strain level to N_2 cycles of another strain level. For example, based on Fig. 4a, it is clear that 100 cycles of $\pm 0.5\%$ strain level will cause a similar amount of strain accumulation to about 19,400 cycles of $\pm 0.2\%$ strain level (follow the vertical strain between 0.5 and 1%). The real challenge is how to understand whether or not a linear strain accumulation model (as in fatigue) is acceptable for soils. In other words, if it is acceptable to take the linear combination of cycles of different strain levels due to extreme events (breaking waves, swell, etc.) throughout the lifetime of the wind turbine according to Eq. (4). An alternative way to avoid these issues is to find out from element tests, a threshold strain where cycles of loading will not cause any strain accumulation. In analogy with Fatigue Limit, this is *Endurance limit* of the material. Zero strain accumulation will suggest no tilting and it may imply larger pile diameter. If zero accumulation is ensured, it will not matter if the loading is one way cyclic or two-way cyclic.

5 Conclusions

A series of element tests using Cyclic Simple Shear (DSS) Apparatus has been carried out to find out the change in shear modulus of the soil under different conditions of cyclic loading. The loading pertained to the particular scenario of offshore wind turbines. The results obtained from the element tests reinforced the observations from the scaled model tests and DEM analysis thereby boosting our confidence in the understanding of the physical mechanism and processes controlling the long-term behaviour of these new structures. While scaled model tests can be insightful to understand the physical mechanisms, the scalability of the results to real application is difficult if the same soil is not used for the model tests. In such cases, element tests provide a better alternative. This paper shows element tests that may be helpful to predict the long-term performance.

References

Abdel-Rahman, Achmus and Kuo. 2014. A numerical model for the simulation of pile capacity degradation under cyclic axial loading. In *8th European conference on numerical methods in geotechnical engineering (NUMGE 2014), Delft*, 18–20 June.

Arany, L., S. Bhattacharya, J. Macdonald, and S.J. Hogan. 2014. Simplified critical bending moment spectra of offshore wind turbine support structures. *Wind Energy*. https://doi.org/10.1002/we.1812.

Arany, Laszlo, S. Bhattacharya, John H.G. Macdonald, and S. John Hogan. 2016. Closed form solution of Eigen frequency of monopile supported offshore wind turbines in deeper waters incorporating stiffness of substructure and SSI. *Soil Dynamics and Earthquake Engineering* 83: 18–32. ISSN 0267-7261, http://dx.doi.org/10.1016/j.soildyn.2015.12.011.

Bhattacharya, S. 2014. Challenges in design of foundations for offshore wind turbines. *IET Journal, Engineering & Technology Reference* 9. https://doi.org/10.1049/etr.2014.0041, online ISSN 2056-4007.

Bhattacharya, S., and S. Adhikari. 2011. Experimental validation of soil–structure interaction of offshore wind turbines. *Soil Dynamics and Earthquake Engineering* 31 (5–6): 805–816.

Bhattacharya, S., D. Lombardi, and D. Muir Wood. 2011. Similitude relationships for physical modelling of monopile-supported offshore wind turbines. *International Journal of Physical Modelling in Geotechnics* 11 (2): 28–68.

Bhattacharya, S., N. Nikitas, J. Garnsey, N.A. Alexander, J. Cox, D. Lombardi, D. Muir Wood, and D.F.T. Nash. 2013. Observed dynamic soil–structure interaction in scale testing of offshore wind turbine foundations. *Soil Dynamics and Earthquake Engineering* 54: 47–60.

Cox, J.A., C.D. O'Loughlin, M. Cassidy, S. Bhattacharya, C. Gaudin, and B. Bienen. 2014. Centrifuge study on the cyclic performance of caissons in sand. *International Journal of Physical Modelling in Geotechnics* 14 (4): 99–115. https://doi.org/10.1680/ijpmg.14.00016.

Hu, Wei-Hua, Sebastian Thöns, Samir Said, and Werner Rücker 2014. Resonance phenomenon in a wind turbine system under operational conditions. In *Proceedings of the 9th international conference on structural dynamics, EURODYN 2014, Porto, Portugal*, 30 June–2 July 2014, eds. A. Cunha, E. Caetano, P. Ribeiro, and G. Müller. ISSN: 2311-9020; ISBN: 978-972-752-165-4.

Leblanc, C. 2009. *Design of offshore wind turbine support structures*. Doctor of Philosophy, Technical University of Denmark.

Lombardi, D., S. Bhattacharya, and D. Muir Wood. 2013. Dynamic soil–structure interaction of monopile supported wind turbines in cohesive soil. *Soil Dynamics and Earthquake Engineering* 49: 165–180.

Nikitas, G., N. Vimalan, and S. Bhattacharya. 2016. An innovative cyclic loading device to study long term performance of offshore wind turbines. *Journal of Soil Dynamics and Earthquake Engineering* 82: 154–160.

Silver, M.L. and H.B. Seed. 1971. Volume changes in sands during cyclic loading. *Journal of Soil Mechanics & Foundations Div.*

Multiple-Driven Fibre-Reinforced Columnar Intrusions for Vertical Drains—A Case Study

Anil Joseph, Babu T. Jose, S. Chandrakaran and N. Sankar

1 Introduction

Kuttanad is an area covering districts of Alapuzha, Kottayam and Pathanamthitta in Kerala. It is well known for its vast paddy fields and geographical peculiarities. Interestingly this region has the lowest altitude in India and farming is carried out at around 1.5–3.0 m below sea level, very much akin to such activities in Netherlands.

The area consists of vast tracts of very soft clay extending to great depths causing several problems while carrying out infrastructural development. Failures of embankments of even 2 m height, when founded on Kuttanad clay is reported (Ayyar 1966). So it is mandatory to improve the ground before going in for any kind of infrastructural development. The effect of techniques such as preloading, stone columns and deep mixed soil—cement column in improving Kuttanad clay have been investigated. (Issac and Girish 2009; Bindu and Vinod 2010; Suganya and Sivapullaiah 2012). The very poor shear strength of Kuttanad clay leads to failure of embankments in general and especially in approach roads of bridges due to greater height of embankment. Rotational failure at toe and base, leads to escape of embankment material to neighbouring areas often encroaching into private lands around. In addition, the longer duration taken for total settlement to occur, makes

A. Joseph (✉)
GeoStructurals (P) Ltd., Cochin, Kerala, India
e-mail: aniljoseph01@gmail.com

B. T. Jose (✉)
Albertian Institute of Science & Technology, Cochin University P.O., Kochi, Kerala, India
e-mail: babutjose@ymail.com

S. Chandrakaran (✉) · N. Sankar (✉)
Department of Civil Engineering, N.I.T. Calicut, Kozhikode, Kerala, India
e-mail: chandra@nitc.ac.in

N. Sankar
e-mail: sankar@nitc.ac.in

K. Ilamparuthi and R. G. Robinson (eds.), *Geotechnical Design and Practice*,
Developments in Geotechnical Engineering,
https://doi.org/10.1007/978-981-13-0505-4_11

the pavement uneven and full of pot holes for initial period which extends to a few years. Faster Consolidation of clay deposits and improvement in its shear strength to withstand the stresses induced by embankment formation are the challenges often faced by geotechnical engineers in this area.

Vertical drains such as Sand piling, PVDs, etc. are often adopted as remedial measures (Indraratna et al. 1994). However, unless the area is preloaded and until the required time for consolidation is allowed, these techniques fail to bring in the desired results and it will not be possible to proceed with construction.

2 Failure of Approach Roads at Bridge Sites

Kerala PWD was constructing a road from Nattakom to Thiruvattukkal in Kottayam District and the area consists of typical deep deposits of Kuttanad clay. The aerial view of the site is shown in Fig. 1. Interestingly the whole stretch of 3 km had been strengthened by PVDs at 0.8–1 m centre to centre in triangular pattern driven to a depth varying from 13 to 25 m depending upon the subsoil profile.

The stretch had two bridges at Parechal and Gramenchira. The approach roads at these locations were 5.8 and 7 m high respectively. In August 2015, the approach road at Gramenchira Bridge settled by up to 2.8 m overnight as shown in Fig. 2.

In the case of Parechal bridge, it was a typical slope failure of embankment resting on very soft clay. The average settlement was around 2.5 m, as shown in Fig. 3.

The ground improvement technique in the form of PVD's did not bring in the desired results as shown by the two failures. Figure 4 shows the failure in the Gramenchira bridge embankment. The authorities had already announced the date for inauguration of the road and bridges which was hardly 5 months away. This brought in severe constraints on project duration. It ruled out further preloading even though the PVDs had already been installed. In addition, the initial shear strength was so poor, any embankment formation, however low it might be, was

Fig. 1 Site plan of the project area

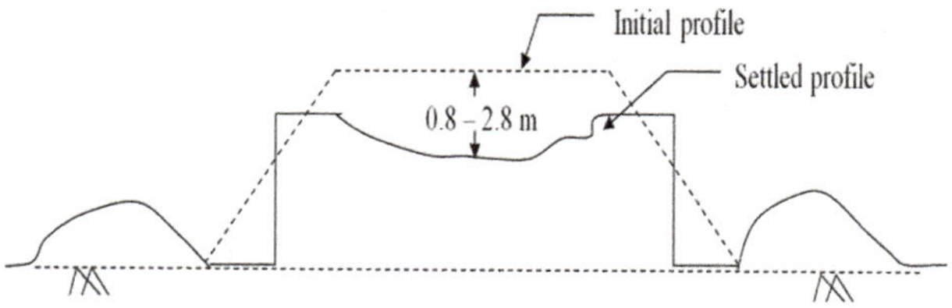

Fig. 2 Failure of approach embankment of Gramenchira Bridge

Fig. 3 Failure of approach embankment of Parechal Bridge

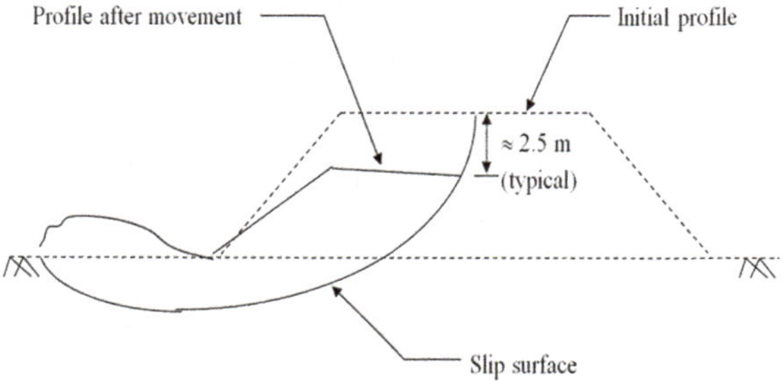

Fig. 4 Failure photographs of Gramenchira Bridge Embankment

always accompanied by patterns of failure as shown in Fig. 2 or 3, especially where embankments were higher.

With hardly 5 months left for inauguration of road and the bridges, the engineers had to look for alternative techniques to overcome the present impasse.

Table 1 Basic properties of soil at Gramenchira and Parechal approaches

S. No.	Properties		Value
1	Natural moisture content,%		100–140
2	Specific gravity		2.67–2.70
3	Grain size distribution	Sand %	0
		Silt %	24–34
		Clay %	66–76
4	Liquid limit,%		110–175
5	Plastic limit,%		45–75
6	Plasticity index,%		65–110
7	Shrinkage limit,%		9–16
8	pH		7.53
9	Coefficient of consolidation (cm^2/sec)		1.48×10^{-4} to 6.83×10^{-4}
10	Compression index C_c		1.07
11	Shear strength from SCPT tests		14 kPa

2.1 *Soil Profile and Properties*

The problem was referred to the authors by PWD. The soil profiles and properties of strata were obtained through fresh soil investigation. The soil exploration programme consisted of five boreholes at Gramenchira and six boreholes at Parechal bridge. In addition Static cone penetration tests were conducted at two locations each in Gramenchira and Parechal. The range of physical and engineering properties of samples collected, obtained from laboratory tests are presented in Table 1.

3 A New Technique Proposed

Discussions considering all parameters involved and deadlines to be met with, led to the adoption of Multiple-driven Fibre-reinforced Columnar Intrusion (MFCI) for ground improvement. This was being tried in certain areas in Cochin and its suburbs for over a decade by the authors. Cochin also is well known for its deep deposits of soft marine clay. Problems of similar nature were often encountered when infrastructural projects were taken up especially along the coastal belt of Cochin.

Various techniques such as sand piling, installation of PVDs, stone columns and numerous other stabilization methods were often tried in such projects which did bring in expected improvement in certain areas. Basic concept of all these innovations was to provide adequate facility for vertical drainage for faster escape of pore pressure. But the improvement could be brought in fully only if the methods are accompanied by preloading for varying periods depending upon the size and spacing of vertical drains. Pre-consolidation is initiated only on preloading which could often be expensive and time-consuming.

Initial attempts consisted of improved version of sand piling, where a second sand pile was driven over and above the first one at the same spot. This forceful intrusion of piling material increased the diameter of normal sand pile and led to development of pore pressure. With the vertical drains already available and pore pressure developed due to multiple driving, the consolidation process was triggered as soon as the multiple driving was completed.

This technology was adopted especially for large oil tank farms right from 2002. Performance of the structures resting on soft clay areas were monitored for over 14 years and their performance has been satisfactory till date. However, laboratory investigations were not taken up simultaneously. But detailed experimental investigations and model studies since 2012 helped to establish the reliability of the new technique and to look for more avenues for its application.

3.1 Laboratory Studies

One of the major limitations of columnar inclusion is the smeared zone generated due to the relative movement of the mandrel and the soil around. This leads to lower capabilities with in the smeared zone. The presence of smear zone significantly influences the horizontal consolidation rate and is reported by various researchers. (Onoue et al. 1991; Hird and Moseley 2000; Basu and Prezzi 2007; Parsa-Pajouh et al. 2014). From the laboratory evaluation, the extension of smear zone for Cochin marine clays is in the range of 5–6 times the radius of the vertical drain. (Joseph et al. 2015). Some of the disadvantages due to smear problems could be overcome by the forceful penetration of the intrusion material into the smeared zone. The experimental setup consisted of a circular tank of 60 cm diameter and 50 cm height (Fig. 5). The tank was made in such a way that it can be removed into five pieces, so that the cross section could be inspected visually and to extract samples from various depths to determine the permeability and strength characteristics in horizontal as well as vertical direction.

After the assembling of the tank it was filled with soft marine clay with water content very close to liquid limit. The mandrel having 10.7 cm in diameter was pressed into this clay at the centre of the tank ensuring that the verticality is maintained throughout the intrusion. River sand was poured into the mandrel and rodded at every 10 cm layer. The mandrel was then slowly with drawn carefully maintaining the verticality.

The diameter of the columnar intrusion was measured. The clay around was carefully troweled off without disturbing the shape of the sand column. The first segment was then detached and the cross section at that level was exposed and measurement of sand column was taken again. This procedure was repeated for every segment of the assembly. Thus the cross section of the vertical sand formation could be visually examined, the shape studied and measurements taken. This

Fig. 5 Procedure of the installation of columnar intrusion and study of configuration on single driving and multiple driving

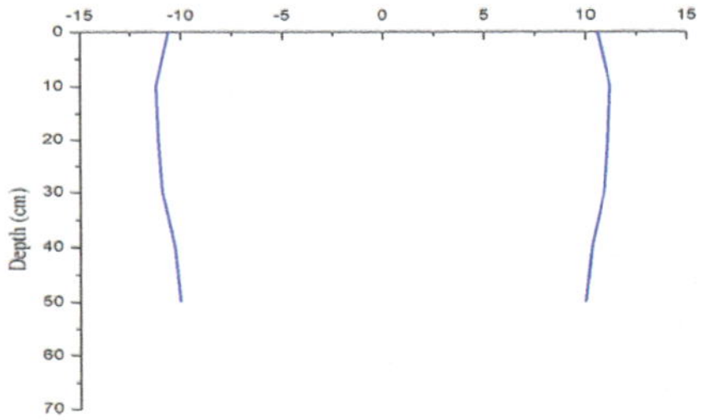

Fig. 6 Diameter of columnar intrusion with single driving (cm)

technique could bring in invaluable information regarding what happens in the soft clay mass during multiple driving. The variation in the diameter of the columnar intrusion noted along the depth for single and multiple driving is given in Figs. 6 and 7. From the comparison of diameter of the columnar intrusion, it is noted that in both the installation the diameter formed are not uniform with the depth. In case of single driving the diameter increases initially indicating bulging of the columnar material and then decreases showing necking characteristics of the soft clay. In case of multiple driving the necking problem is overcome due to the piercing of the columnar intrusion into the clay and considerable intrusion of the material is noted into the smear zone. It is observed that the second drive helps to further densify the material of columnar intrusion which will bring in greater stability to the columns and the results show that the multiple driving processes has pronounced influence on the smear zone.

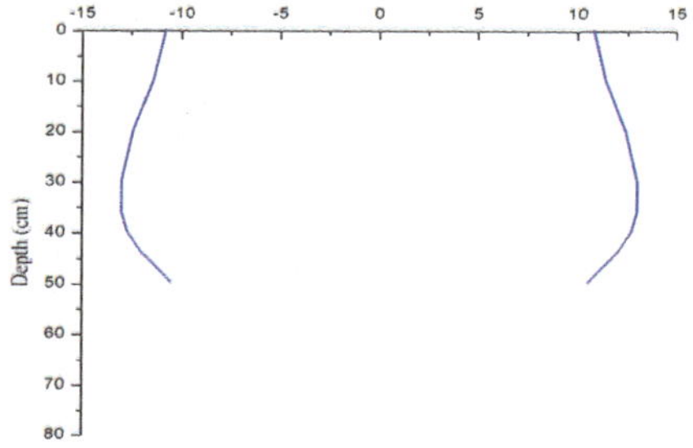

Fig. 7 Diameter of columnar intrusion with multiple driving (cm)

3.2 Material for MFCI

It is found from laboratory experiments that in case of very soft clay, if sand/metal alone is provided during sand piling, the diameter of the well gets reduced due to necking action and the clogging of well is likely to be high. In order to reduce these effects and to increase the strength and permeability characteristics of the drainage media a hybrid mix comprising of metal of varying grades, M sand and fibre reinforcements such as Recron 3S is being evolved through experimental model studies.

One of the major limitations of sand piling and PVD's is the smear zone generated due to the relative movement of the mandrel and the soil around. This leads to lower permeability within the smear zone. In addition, due to the surface tension formed on withdrawal of the mandrel, there will not be direct contact of the columnar material with the soft clay and unless a preload is applied drainage path to dissipate the pore pressure formed in the clay will not be fully effective. It is noted that the forceful penetration of the MFCI material during the second and third driving at the same spot will decrease the loss in permeability in the smear zones, overcome the components of surface tension thereby commencing the dissipation of pore pressure without preloading and reduce the well resistance.

4 Field Application

Considering the site conditions and the tight completion schedules it was decided that, Multiple-driven Fibre-reinforced Columnar Intrusions may be used to stabilize the foundation of the approach roads. These comprise of small diameter (15 cm) columns of specified sand-aggregate mix reinforced with fibres installed to the designated depth (800 cm) and at designed spacing (80 cm, triangular spacing).

These are installed by driving a casing pipe with a flap valve or sacrificial shoe at the bottom and then filling it with a fibre-reinforced sand-aggregate mix. Subsequently, casing is again driven through the installed column and again filled with fibre-reinforced sand-aggregate mix. The process is continued till sufficient

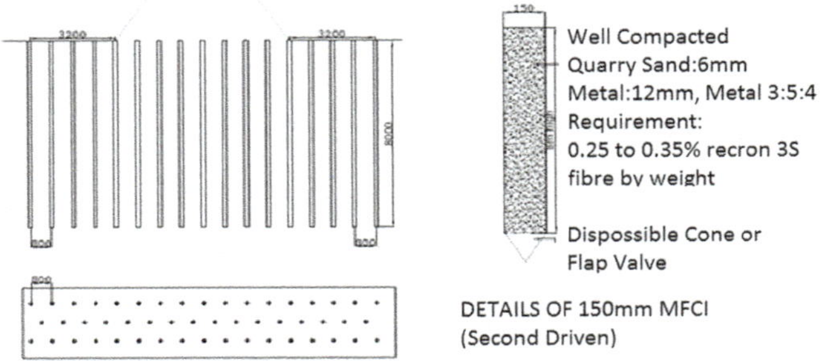

Fig. 8 Layout and sectional plan for implementation of MFCI in bridge approaches at Gramenchira and Parechal

Fig. 9 Implementation of MFCI in bridge approaches at Gramenchira

volume of material is fed into the area and proper compaction is achieved. Figure 8 shows the details of MFCI materials and the specifications used for the current project. After completion of the installation of the intrusions, a 50 cm thick drainage blanket comprising clean aggregates was provided over the entire area improved by MFCI. Over this, the embankment was constructed by placing and compacting in layers (Fig. 9).

4.1 Field Instrumentation and Results

In order to evaluate the performance of the multiple fibre-reinforced columnar intrusions in stabilizing the embankment, field measurements were carried out regularly by taking measurements of vertical settlement with the help of settlement gauges and horizontal movements of the embankment fill with inclinometers.

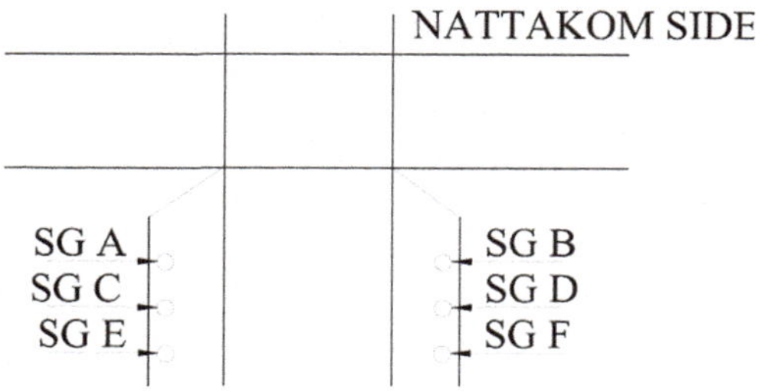

Fig. 10 Location of settlement gauges at Gramenchira

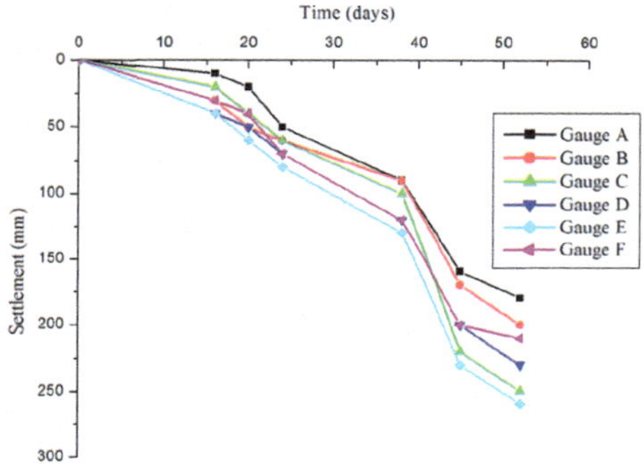

Fig. 11 Time-settlements curves from settlement gauges after MFCI application

The measurement of settlement using settlement plates was done at six locations in field for each embankments near Gramenchira Bridge as shown in Fig. 10.

Figure 11 shows the variation of settlement with time noted for the Gramenchira embankment approach after the ground improvement using the multiple fibre-reinforced columnar intrusion. It is noted that the average settlement of the embankment over two months is 22 cm.

The field engineers started formation of the embankment within a couple of weeks after MFCI installation was completed. However, no sign of distress was in evidence indicating faster gain in strength and resilience obtained from the ground improvement technique adopted. Figure 12 shows the variation noted in the shear strength with depth from the SCPT tests conducted at Gramenchira embankment approach before and after the ground improvement using the multiple fibre-reinforced columnar intrusion. From the comparison of shear strength of soil at area improved with columnar intrusion and in area where no improvement was done a variation of around 65–75% was observed. The average shear strength of 0.14 kg/cm^2 was improved to 0.23 kg/cm^2.

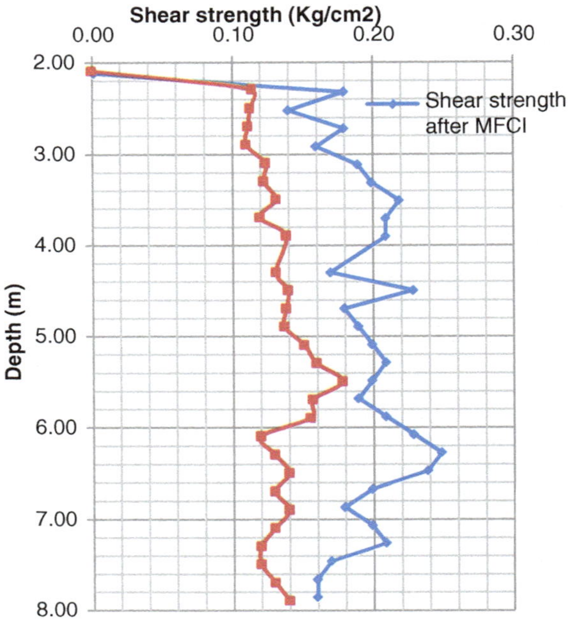

Fig. 12 Variation of shear strength with depth before and after improvement using MFCI technique at Gramenchira bridge approach

5 Conclusions

Sand piling and use of PVD's are accepted practices for ground improvement. Both help to cut short the drainage path and allow faster dissipation of pore pressure. However both require preloading to initiate pre-consolidation and improve the strength characteristics. Smear effect generated by the relative movement of mandrel and soil around and consequent reduction in permeability is a major limitation in vertical drains. It has been overcome to a great extent by penetration of MFCI material into the soil.

In the present case study, even though the PVD's were installed for more than a year, the effect of this either on pre-consolidation or on gain in shear strength was limited. These ultimately led to the failures of approach roads at Gramenchira and Parechal Bridges. Thus where shorter project duration is a prime consideration, the sand piling and the PVD methods may not provide the desired solutions.

MFCI can yield reliable results for ground improvement. The rate of settlements of the improved embankments is monitored through proper instrumentation and the performance has been satisfactory. Further SCPT test conducted before and after application of MFCI shown enhancement of shear strength by 23% within 2 months. These results underscores the acceptability of the proposed techniques. Commencement of embankment formation within days of MFCI treatment did not show any distress indicating faster gaining strength which was verified in field using static cone penetration test. These results underscores the utility and versatility of the new technique.

References

Basu, D., and M. Prezzi. 2007. Effect of smear and transition zones around prefabricated vertical drains installed in a triangular pattern on the rate of soil consolidation. *International Journal of Geomechanics* 7 (1): 34–43.

Bindu, J., and P. Vinod. 2010. Design of preload for reducing the compressibility of kuttanad clays. *Proceedings of International Conference on Materials, Mechanics and Management*, 1: 424–428, (IMMM-2010).

Hird, C.C., and V.J. Moseley. 2000. Model study of seepage in smear zones around vertical drains in layered soil. *Geotechnique* 50 (1): 89–91.

Indraratna, B., A.S. Balasubramaniam, and P. Ratnayake. 1994. Performance of embankment stabilized with vertical drains on soft clay. *Journal of Geotechnical Engineering, ASCE* 120 (2): 257–273.

Isaac, D.S., and M.S. Girish. 2009. Suitability of different materials for stone column construction. *Electronic Journal of Geotechnical Engineering* 14: 1–12.

Joseph, Anil, S. Chandrakaran, N. Sankar, and Babu T. Jose. 2015. Laboratory evaluation of extent of smear zone due to columnar intrusion for cochin marine clays. In *50th Indian Geotechnical Conference, December, 2015*, Pune, Maharashtra, India.

Onoue, A., N.H. Ting, J.T. Germaine, and R.V. Whitman. 1991. Permeability of disturbed zone around vertical drains. In *Proceedings of the geotechnical engineering congress (GSP 27)*, 879–890. New York: American Society of Civil Engineers.

Parsa-Pajouh, Ali, Behzad Fatahi, Philippe Vincent, and Hadi Khabbaz. 2014. *Trial embankment analysis to predict smear zone characteristics induced by prefabricated vertical drain installation*. Springer International Publishing Switzerland.

Ramanatha Ayyar, T.S. 1996. *Strength characteristics of kuttanad clays*. Ph.D. thesis, University of Roorkee, Roorkee.

Suganya, K., and P.V. Sivapullaiah. 2012. Parametric study of embankments founded on soft organic clay using numerical simulations. In *ISSMGE—TC 211 international symposium on ground improvement IS-GI Brussels*, 31 May–1 June 2012.

Development of Fly Ash Stabilized Recycled Base Material (FRB) for Indian Highways

Sireesh Saride and Deepti Avirneni

1 Introduction and Background

Recycling is one of the key components for sustainable development of infrastructure. However, recycled materials such as construction and demolition waste from pavements and byproduct from thermal power plants are highly produced, which are either stockpiled or landfilled across the world if not utilized in large quantities. Reuse of these secondary materials in civil engineering applications in large quantities is always a challenge due to their inferior properties. Safe disposal of recycled construction materials and industrial by products such as RAP, fly ash, waste tires, quarry waste, construction, and demolition waste, etc., has become a real environmental challenge. RAP is a reclaimed material obtained when a portion of distressed asphalt pavement is milled. RAP has been a resource material, when properly stabilized, for a new surface or a base course layer of flexible pavements (Taha et al. 2007). The strength and stiffness properties of RAP can be enhanced by stabilizing them with cement/lime/fly ash (Mohammadinia et al. 2014). Fly ash being another industrial byproduct, if adopted as a stabilizer, can address the land filling concerns. However, the strength, stiffness and durability of these mixes are questionable due to the presence of the thin asphalt coat present over the RAP aggregates (Saride et al. 2016; Avineni et al. 2016). To enhance the strength and stiffness properties of fly ash stabilized RAP bases, alkali activators are adopted (Saride et al. 2016).

Durability in this context may be defined as the ability of the mix to resist severe moisture and temperature fluctuations, chemical attack, and abrasion while

S. Saride (✉)
Civil Engineering Department, Indian Institute of Technology Hyderabad, Hyderabad, India
e-mail: sireesh@iith.ac.in

D. Avirneni
Civil Engineering Department, VNRVJIET, Hyderabad, India

K. Ilamparuthi and R. G. Robinson (eds.), *Geotechnical Design and Practice*,
Developments in Geotechnical Engineering,
https://doi.org/10.1007/978-981-13-0505-4_12

sustaining its desired engineering properties. The durability of cement treated VA: RAP mixes was studied by Ganne (2009) by subjecting the mixes to alternate wet/ dry cycles to measure the resistance to weight loss after 14 cycles. Ganne (2009) found that the percentage strength loss was around 10–15% on an average for all the mixes studied. It is also important to study the permanency of the design mixes as few researchers have reported the loss of stabilizers from the base layers after certain service period. These problems are due to the detrimental secondary chemical changes in the stabilization process.

In the current study, to develop a sustainable fly ash stabilized recycled base (FRB) material, initially, the influence of alkali-activated fly ash stabilization on the strength development of the RAP:VA mixes is evaluated in terms of the unconfined compressive strength. Then the strength of the design mixes after performing the durability and permanency studies on 28-day cured specimens are evaluated. The interaction of activated fly ash with RAP:VA mixes in the strength gain is presented through Fourier transform infrared spectroscopic analysis and X-ray diffraction studies.

2 Materials Used

The RAP was collected from an ongoing cold-milling operations at the National Highway No. 5 at Nellore, Andhra Pradesh. The grain size distribution of the RAP was determined according to ASTM C 136 (2014) and the gradation curve is presented in Fig. 1. As the present gradation of RAP is deviating from the cement stabilized base course material according to the Ministry of Road Transportation and Highways (MoRTH 2000), the material was segregated and maintained an

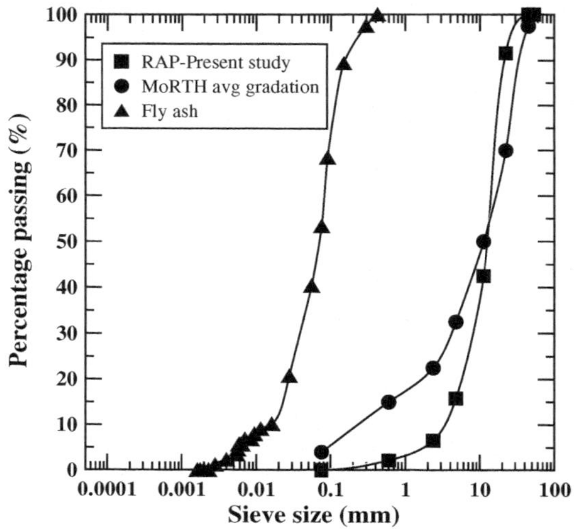

Fig. 1 Grain size distribution of RAP and FA

average grade proposed by the agency. The average grade Virgin aggregates (VA) collected from nearby quarry. The specific gravity of the RAP and VA are found to be 2.7 and 2.6 respectively.

Fly ash was collected from Neyveli Lignite Corporation Limited (NLC), Tamil Nadu. Chemical composition of the fly ash was obtained from X-Ray Florescence (XRF) method. The calcium oxide (CaO) content is about 12% and according to ASTM C 618 (2012a), the fly ash can be classified as class "F" as CaO is less than 20%. The gradation curve of the fly ash is presented in Fig. 1. The specific gravity of the fly ash was found to be 2.21.

The laboratory grade sodium hydroxide pellets with 98.9% purity was used as an activator. Since, the reaction between NaOH and water is exothermic, utmost care should be taken during the preparation of the solution. The solution shall be prepared at least 8 h prior to the mixing.

3 Methodology

According to the Indian Roads Congress (IRC: 37-2012), cement stabilized bases after 28-day curing period should exhibit a minimum unconfined compressive strength (UCS) of 4.5 MPa to qualify as base/subbase material. To meet this requirement, UCS tests were conducted on the specimens prepared and cured in moist chamber at a constant temperature (25 °C) and humidity (75% relative humidity) for 28 days. The RAP:VA mix proportions of 80:20 and 60:40 stabilized with 20 and 30% fly ash were considered in the present study with and without 2 and 4% NaOH activation. The fly ash dosages were adopted based on the prior performance of the fly ash stabilized RAP:VA mixes reported by the authors (Saride et al. 2015). The mix proportions and their nomenclature followed throughout the paper are shown in Table 1. Prior to the UC strength tests, modified Proctor compaction tests were carried out in accordance with ASTM D 1557-12 (2012b) for different mixes to determine optimum moisture content (OMC) and maximum dry density (MDD). Table 1 presents the compaction characteristics of the mixes. With an increase in the dosage of fly ash, as expected, the compaction curves have shifted towards the right hand side, reflecting an increase in the OMC.

Table 1 Mix proportions and their nomenclature

RAP:VA	FA	N	Nomenclature	OMC (%)	MDD (g/cc)
80:20	20	0	80R:20A + 20F + 0 N	7.05	2.16
80:20	30	0	80R:20A + 30F + 0 N	7.25	2.14
80:20	20	2	80R:20A + 20F + 2 N	7.00	2.17
80:20	30	2	80R:20A + 30F + 2 N	7.20	2.15
80:20	20	4	80R:20A + 20F + 4 N	6.90	2.17
80:20	30	4	80R:20A + 30F + 4 N	7.15	2.18

The UC strength tests were also conducted after a repeated wet/dry cycles (durability) on compacted specimens of size 100mm × 200 mm. The procedure outlined in ASTM D 559 (1996) was followed to simulate the moisture fluctuations due to seasonal variation in the field. According to ASTM D 559 (1996), at the end of 28-day curing, the specimens were immersed in potable water for a duration of 5 h at room temperature. The samples were then removed and oven dried at 70 °C for 42 h. This exposure of specimens for 47 h to both wetting and drying activities constitutes a single cycle. Then, the specimens were weighed. The procedure is repeated for 12 such cycles and each specimen is weighed after each cycle to calculate the weight loss.

Apart from the seasonal variations, water infiltration during rainfall may also alter the performance of these mixes. Hence, leachate or permanency studies were performed to verify the loss of stabilizer due to the rain water infiltration and its impact on the UC strength. Any loss in the reactants can be observed by testing the leachate for the respective ion concentrations and pH variation. In these tests, 28-day cured specimens were subjected to water percolation under a controlled pressure of 50 kPa, such that, the packing of the sample is not disturbed. The pore volume of the compacted specimens was calculated and the collection of in-filtered water equivalent to each pore volume constitutes one cycles. Leachate tests were continued until such 14 pore volumes of in-filtered water are collected.

To verify and quantify the formation of pozzolanic compounds, mineralogical studies such as X-ray diffraction (XRD) studies and Fourier transform infrared spectroscopic analysis (FT-IR) were conducted.

4 Results and Discussion

4.1 UC Strength and Durability

The results from UCS tests conducted on fly ash treated 80R:20VA and 60R:40VA mixes with and without 2 and 4% NaOH activation after the performance of durability studies are presented in Fig. 2. The results indicate that, at all fly ash dosages, the mixes without alkali activation (mixes 1–4) do not satisfy the minimum UCS requirement of 4.5 MPa as per the IRC: 37-2012 at 28 day curing. At 2 and 4% NaOH, except 80R:20 V + 20F with 2% NaOH specimens (mix 5), all the other mixes (mixes 6–9) have satisfied the specified strength requirement. The increase in UC strength is observed to be about 1.7–1.8 times when activated with 2% NaOH. This increase in strength clearly indicates the enhanced reactivity of the fly ash due to alkali activation. In addition, with the increase in NaOH dosage from 2 to 4%, all the mixes have gained about 10% of additional strength.

During the durability studies, the weight loss of all the mixes is observed to be well within the permissible limit of 14% as specified in the IRC: 37-2012 for a cemented base material. It was observed that for a given RAP proportion (80%), the

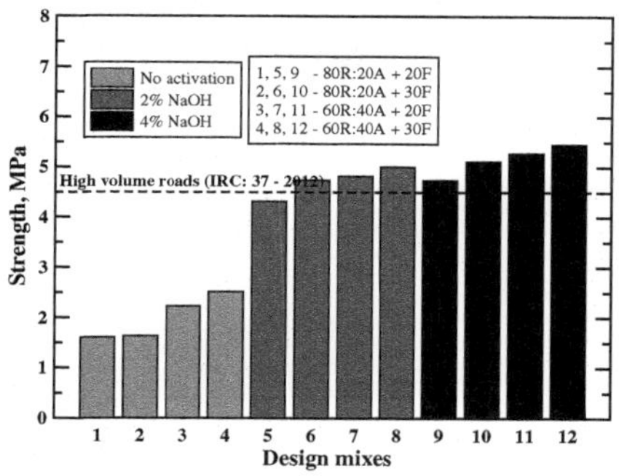

Fig. 2 Variation of UCS after 12 wet-dry durability cycles for different design mixes

Table 2 Percentage of Ca and Na leached out with respect to the initial concentrations mixed

Ion	Initial concentration, molarity (ppm)	Cumulative concentration after 14 leachate cycles (ppm)	Percentage leached (%)
Ca	240,000	98	0.04
	300,000	214	0.07
Na	23,000	1350	5.86
	46,000	1592	3.46

weight loss is higher for low fly ash (20%) and NaOH (2%) contents. This may be attributed to the hindered pozzolanic activity due to the presence of high amount of aged bitumen in the mix and relatively low cementation. In addition, during the permanency or leachate studies, the leaching of the stabilizer and activator were recorded as shown in Table 2. It is found that about 50 ppm of CaO leached out of the specimens during the first cycle, thought in the permissible limits, becomes less than 15 ppm in fourth cycle, indicating that the excess lime present in the mix has leached out. On an average, the pH concentration of the leachate was observed to be reduced from 12.5 to 9.5 in 14 cycles. The initial high pH concentration of the leachate can be attributed to the amount of free CaO liberated from the mixes. Moreover, the percentage of leached calcium and sodium ion concentration is about 0.07 and 5.9% respectively, which indicates the permanency of the stabilizers in the treated sections.

4.2 Resilient Modulus

In addition to the UCS tests, resilient modulus (M_r) of RAP:VA mixes with 20 and 30% fly ash specimens were determined in accordance with the American Association of State Highway and Transportation Officials, AASHTO T-307

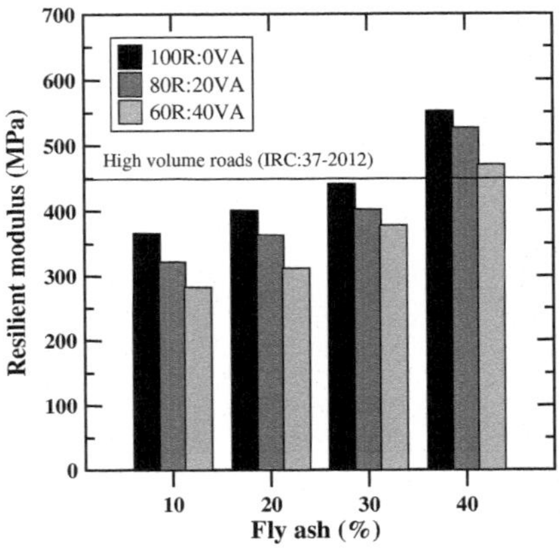

Fig. 3 Resilient modulus of RAP:VA mixes with different dosages of fly ash

(AASHTO 2003). The M_r is defined as the ratio of cyclic deviatoric stress to the recoverable strain. The resilient modulus tests were also performed on 100 percent RAP specimens along with 10 and 40% fly ash dosages. The entire data is necessary to develop design charts. It can be noticed that an increase in the VA content in the mix decreased the resilient modulus by about 5 and 15%, respectively for 80R:20VA + 40F mix and 60R:40VA + 40F mix (Fig. 3). It is due to the superior resilient behavior of the RAP material to VA as RAP has already undergone certain number of load repetitions in the form of wheel loads in its life span. This observation is in conjunction with the observations made by King (2001).

4.3 Mineralogical Studies

The X-ray diffraction studies are used to identify the crystalline phase of the materials. As the products of the pozzolanic reactions are of crystalline in nature, in the present study, the influence of the alkali activation on fly ash treated RAP mixes was analyzed using XRD. The influence of RAP dosage on the hydration and cementation in the mix is presented in Fig. 4. It is observed that the mix with high

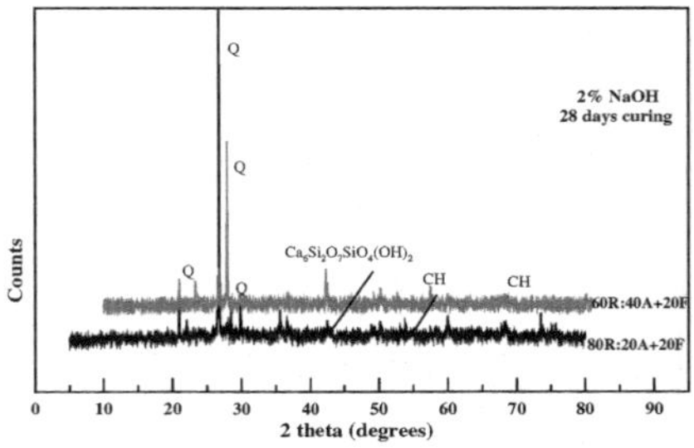

Fig. 4 XRD profiles of different mixes with 2% NaOH at 28 day curing

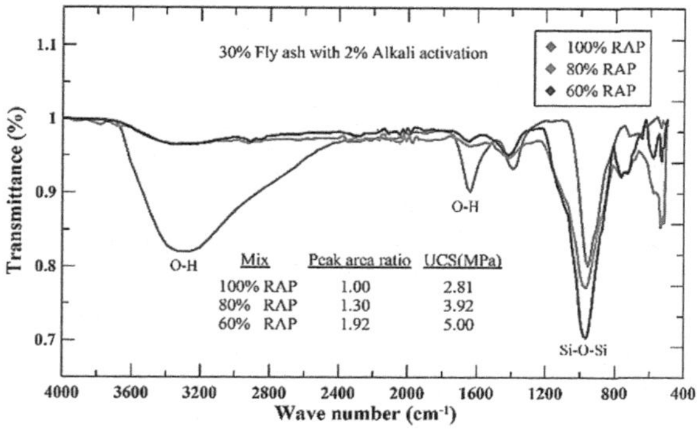

Fig. 5 Typical FT-IR spectra of 80R:20VA + 30F with and without activation at 28 day curing

RAP content exhibited lower cementation because of its inert nature. It is to be noted that $Ca(OH)_2$ is still available for further reactions even at 28 days curing. Hence, further reactions are expected with the increase in curing period which will eventually improve the strength of the mix.

The FT-IR spectra corresponding to 100, 80, and 60% RAP stabilized with 30% fly ash and activated with 2% NaOH at 28 days curing is presented in Fig. 5. It can be observed that the intensity of O–H band is high for the mix containing 100% RAP. This is because RAP is less prone to water absorption and reactivity. Figure 5 also shows the area of peak heights of Si–O–Si band. The peak area and peak height ratio were highest for the mix containing 60% RAP, indicating best degree of hydration which can also be supported with UC strength results. The bands appearing between 1100 and 900 cm^{-1} are due to the asymmetric stretching vibrations of Si–O–Si or Al–O–Si. This is referred to the formation of amorphous to semi-crystalline alumino-silicate materials.

Interesting observation is that this band shifted from 1050 to 950 with the addition of NaOH, which can be attributed to the formation of hydration products. It is reported that the peak areas and peak heights at Si–O–Si stretching vibrations cab be used for the quantitative assessment of the reactions (Lodeiro et al. 2009). The peak areas are calculated using SCION, an image processing software which gives area of the selected boundary in pixels. The ratios of peak area of Si–O–Si stretching vibration of the mixes are also presented in Figs. 5 and 6. The values correspond to the normalization with respect to the mix without activation. For the mix without activation, peak area ratio is relatively low. The peak area ratio of the mix increases with an increase in the activator content from 0 to 4% suggesting that a relatively high degree of hydration has occurred. These results are also in accordance with the trends observed in UC strengths which are presented in Fig. 6.

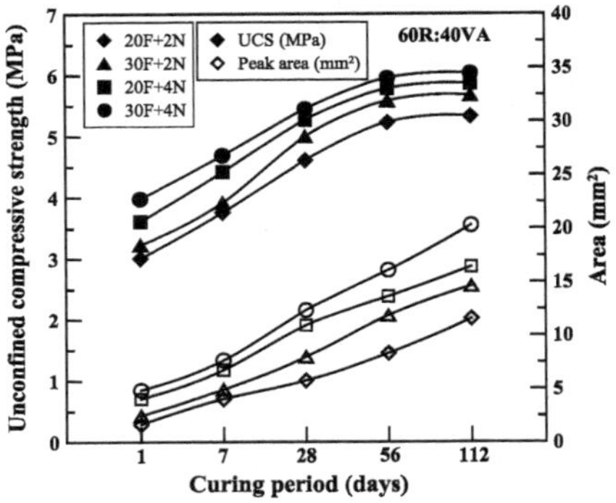

Fig. 6 Variation of UCS and peak areas with curing periods for 60R:40VA mixes

5 Design Charts

An attempt has been made in the present study to develop a design chart for the traffic in terms of standard axels varied from 2 to 150 msa (million standard axles) which covers a wide range of traffic options. A weak subgrade material (with CBR 3%) is considered. The elastic modulus of the subgrade is calculated as per Eq. (1).

$$E(\text{MPa}) = 10 \times \text{CBR for CBR} \leq 5 \tag{1}$$

The resilient modulus of base/subbase material is varied between 100 and 600 MPa, which is a practical range for low-grade recycled materials. If the cementitious materials are used in the base/subbase layers, an aggregate interlayer of 100 mm thick unbound granular layer with a modulus value of 450 MPa should be provided which acts as a crack relief layer as per IRC (2012).

The elastic modulus of bituminous mix is highly temperature-dependent and also depends on the grade of the bitumen used. Hence, the design is performed for a modulus value of 2500 MPa. A modulus of 2500 MPa can be achieved for a penetration binder grade of 60–70 at 30 °C. Poisson's ratio of bituminous and base layers is considered as 0.35 and for the weak subgrade as 0.4. The typical design chart for various traffic conditions is presented in Fig. 7.

A reduction in base layer thickness can be obtained if the modulus of the surface layer is increased from 2500 to 3000 MPa. If the CBR of the subgrade soil is increased, a considerable reduction in base layer thickness can be obtained. For a heavy traffic of 150 msa, the resilient modulus of base material should be at least 300 MPa is preferred.

The advantages of the proposed design chart are summarised as follows.

- It is a very simple method: Once the M_r value of the base/subbase material is known, the design takes only few minutes to arrive at the design thickness.
- It is resilient modulus based design, and hence it is more accurate.
- It considers a range of M_r values which will be more realistic for recycled and locally available low grade materials (with M_r < 600 MPa).

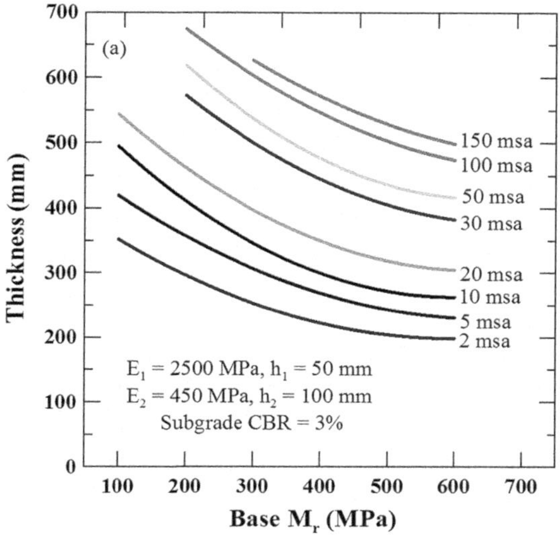

Fig. 7 Design chart for surface layer M_r of 2500

- These design charts can be further extended to stabilized base course materials viz. cemented RAP, cemented VA and any other suitable stabilized materials, whose resilient modulus values are expected to range from 450 to 3000 MPa.

6 Conclusions

The following conclusions can be drawn from this paper:

In spite of the presence of excess calcium, fly ash stabilization could not impart sufficient UC strength to the RAP:VA mixes. Hence, fly ash activation is necessitated.

Around threefold increase in strength for all the mixes was observed with NaOH activation. Mixes activated with 4% NaOH showed 15% higher improvement in strength compared to the mixes with 2% NaOH activation.

The fly ash was successfully activated with 2 and 4% NaOH to further the strength gain to meet the design strength 4.5 MPa.

The durability and permanency studies in terms of aggressive wet/dry cycles indicated that the activated fly ash stabilized RAP:VA mixes performed satisfactorily. The weight loss after the durability cycles was found to be less than 14% as specified by IRC: 37-2012.

Reduction in intensity of bands formed due to free—OH groups ($n1$) of water molecules with curing period indicate that the hydration is proceeding. The asymmetric stretching vibrations of Si–O–Si increased with activator content and curing period thereby implying the formation of hydration products which is also confirmed by the increased UC strength.

A design chart is proposed to adopt the FRB as a potential base course mix for sustainable Indian highways.

Overall, it is observed that the geopolymer fly ash treated RAP:VA base course mixes can be successfully employed in new pavement construction.

Acknowledgements Authors would like to thank the Technology Systems Development program of Department of Science and Technology, government of India for providing financial assistance to carry out this project (Project Sanction No. DST/TSG/STS/2013/40-G). Authors would like to thank M/S Neyveli Lignite Corporation Limited for their support by supplying fly ash for this research study.

References

AASHTO T-307. 2003. Standard method of test for determining the resilient modulus of soils and aggregate materials. Washington, DC.

ASTM. 1996. Standard test methods for wetting and drying compacted soil-cement mixtures. West Conshohocken, PA: ASTM D559-96, ASTM International, 1996. www.astm.org.

ASTM. 2012a. Standard specification for coal fly ash and raw or calcined natural pozzolan for use in concrete. West Conshohocken, PA: ASTM C 618-12a.

ASTM. 2012b. Standard test methods for laboratory compaction characteristics of soil using modified effort (56,000 ft-lbf/ft^3 (2,700 kN-m/m^3)). West Conshohocken, PA: ASTM D1557-12e1 ASTM International. www.astm.org.

ASTM. 2014. Standard test method for sieve analysis of fine and coarse aggregates. West Conshohocken, PA: ASTM C136 (2014).

Avirneni, D., P.R.T. Peddinti, and S. Saride. 2016. Durability and long term performance of geopolymer stabilized reclaimed asphalt pavement base courses. *Construction and Building Materials* 121: 198–209.

Ganne, V.K. 2009. *Long-term durability studies on chemically treated reclaimed asphalt pavement materials*. Master thesis report. University of Texas, Arlington.

Indian Roads Congress. 2012. Guidelines for the design of flexible pavements. New Delhi, India: IRC-37.

King, B. 2001. Evaluation of stone/rap interlayers under accelerated loading construction report. Report no. FHWA/LA-352, Baton Rouge, LA.

Lodeiro, G., D.E. Macphee, A. Palomo, and A.F. Jimenez. 2009. Effect of alkalis on fresh C–S–H gels. FTIR analysis. *Cement and Concrete Research* 39: 147–153.

Mohammadinia, A., A. Arulrajah, J. Sanjayan, M. Disfani, M. Bo, and S. Darmawan. 2014. Laboratory evaluation of the use of cement-treated construction and demolition materials in pavement base and subbase applications. *Journal of Materials in Civil Engineering* 27 (6): 04014186. https://doi.org/10.1061/(ASCE)MT.1943-5533.0001148.

MoRTH. 2000. *Specifications for road and bridge works*. Published by the Indian Roads Congress on behalf of the Government of India, Ministry of Road Transport and Highways, Government of India.

Saride, S., D. Avirneni, and S. Javvadi. 2015. Utilization of reclaimed asphalt pavements in indian low-volume roads. *Journal of Materials in Civil Engineering* 28 (2): 04015107. https://doi.org/10.1061/(ASCE)MT.1943-5533.0001374.

Saride, S., D. Avirneni, and Ch. Subrahmanyam. 2016. Micro-mechanical interactions of activated fly ash mortar and reclaimed asphalt pavement materials. *Construction and Building Materials* 123: 424–435.

Taha, R., G. Ali, A. Basma, and O. Al-Turk. 2007. Evaluation of reclaimed asphalt pavement aggregate in road bases and sub-bases. *Transportation Research Record* 1652: 264–269.

Impact of Anisotropy in Permeability of Peaty Soil on Deep Excavation

D. E. L. Ong

1 Case Study 1

1.1 Introduction

The Kuching Wastewater Management System (KWMS) Phase 1 was implemented to channel 100,000 population-equivalent household grey and blackwater via a full-gravity underground network of sewer pipelines to a centralised wastewater treatment plant. To facilitate sewer pipeline construction via pipe-jacking method, shafts with depths of between 10 and 25 m were constructed to enable the launch and reception of micro-tunnelling boring machines. The construction of shafts in the Tuang Formation posed great challenges especially in sandy and organic soils as appreciable ground water drawdown could be resulted, followed by the detrimental effects of ground settlements.

1.2 Geology

The geology of the area is characterised by deltaic deposits consisting peat, clay, silt and sand layers overlying Kuching's Tuang Formation. The current project site was located at a low-lying area where sediments were deposited due to receding sea levels in historical events. The sediments remained trapped at foothills and organic matters started to accumulate to form rather horizontal beds of clays or silts with organic matters. Figure 1 shows the texture of peaty soil found at the project site. Over long

D. E. L. Ong (✉)
Faculty of Engineering, Science & Computing, Research Centre for Sustainable Technologies, Swinburne University of Technology, Sarawak Campus, 93350 Kuching, Sarawak, Malaysia
e-mail: elong@swinburne.edu.my

K. Ilamparuthi and R. G. Robinson (eds.), *Geotechnical Design and Practice*, Developments in Geotechnical Engineering,
https://doi.org/10.1007/978-981-13-0505-4_13

Fig. 1 Texture of peaty soil

(a) (b)

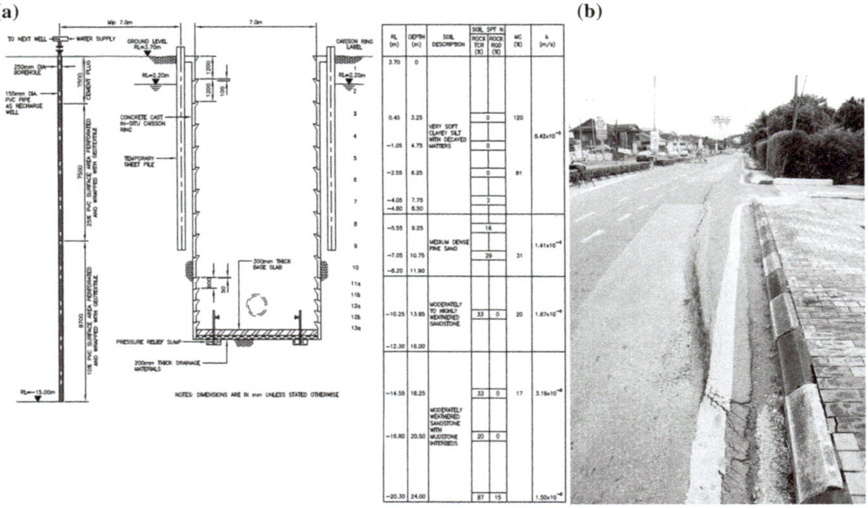

Fig. 2 **a** Typical deep shaft construction and location of large-diameter recharge well and **b** 125 mm settlement caused by ground water drawdown

periods of time, the surface organic deposits rose and finally remained above the flood level. The underlying Tuang Formation rock has been described as 'erratic' by Ong and Choo (2011, 2012) in view of its highly sheared and weathered rock mass.

1.3 Details of Deep Shaft

Figure 2a shows the details of the proposed 7 m diameter, 16 m deep shaft. Sheet piles were first driven around the shaft in an attempt to lengthen the ground water flow path in order to reduce the risk of floor uplift due to the difference in hydrostatic pressure inside and outside the shaft that was being excavated. Advancing cast in situ circular concrete caisson rings (each measuring 1.3 m in height, with 100 mm overlap) were then installed to provide structural support to the surrounding soils so that excavation inside the shaft could be carried out.

In this method, the concrete caisson rings served as temporary works and the design was mainly based on hoop stresses generated by lateral soil pressures. As such, the steel reinforcement used was nominal since the hoop stresses were designed to be taken entirely by the concrete.

1.4 Effect of Leakage at Underlying Fractured Rock

At one of the many challenging excavation sites, when the 7 m diameter, 16 m deep shaft that was being excavated reached the moderately to highly weathered sandstone mass at depth of about 13.3 m, major leakage occurred. Hence, the deep excavation works had to be suspended. Consequently and inevitably, ground water drawdown of about 5 m occurred, triggering settlement of 125 mm to nearby infrastructure as shown in Fig. 2b, mainly due to the relatively high permeability of the 8.5 m upper soft, organic soils with SPT $N = 0$.

1.5 Ground Water Responses

The quantity and rate of the ground water that seeped into the leaking shaft could be estimated on site by simply calculating upwards from the base of the shaft the number of caisson rings flooded in the shaft after it was successfully pumped dry.

The amount of water that rose in the shaft overnight (no pumping activity) would then give a reliable estimate of seepage rate into the shaft. In this case, the 7.0 m diameter shaft simply acted like an oversized water standpipe.

The volumes of water and flow rates vary on different days because the exposed rock faces had different characteristics and permeability values as the contractor excavated into the rock mass over time with an average seepage rate of 13.6 m^3/h.

1.6 Fem

On average, the measured seepage was 13.6 m^3/h, which was reasonably comparable to the estimated ground water seepage rate of 17.4 m^3/h based on the finite element seepage analyses conducted during the design stage (see Fig. 3).

In order to reflect the actual soil and sheet pile wall interaction behaviour, slip interface has been modelled with realistic values. Interface elements or slip

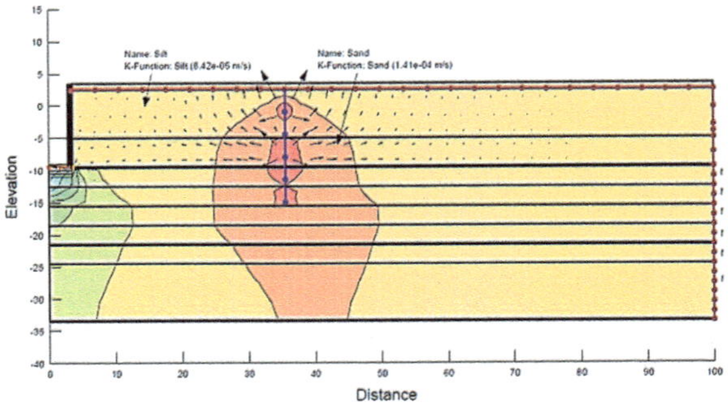

Fig. 3 Typical transient seepage analysis showing total pressure head contours

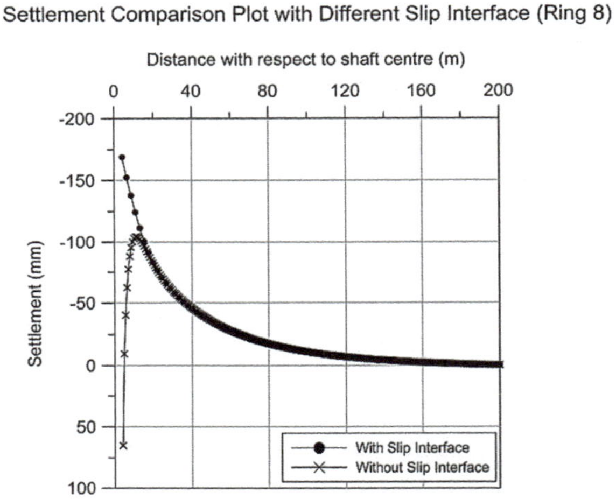

Fig. 4 Settlement profile with and without the use of interface slip elements

elements are used in finite element analyses to simulate sliding between two different materials. These elements have thin width and shear stiffness comparable or less than the surrounding materials. The interface poisson ratio (ν_i) of 0.45, interface friction angle of 14° or equivalent to friction factor of 0.25 have been adopted and calculated average interface shear modulus (G_i) along the sheet pile wall and soil interface which is 249.72 kPa.

Figure 4 shows ground surface settlement at excavation depth of 9.7 m or Ring 8 with or without slip interface and with different G_i and 2 times of G_i. With higher values of interface shear modulus, the ground settlement curve experiences less ground surface settlement next to the sheet pile wall region. For the case with no slip interface, the ground surface settlement curve shows a clear dragging effect between the interface of sheet pile wall and soil.

1.7 Ground Treatment Works

As a form of ground treatment works, large-diameter recharging wells were installed to recharge the depressed ground water table at site. Field monitoring showed that ground settlements of not more than 7.5 mm were observed when the excavation works resumed. It took a further 17 days to complete the remaining excavation works, which consisted merely a further excavation of 1.8 m and casting of the base slab. Pressure relief sumps, control valves as well as lined drainage materials were constructed at the shaft floor to control water seepage coming into the shaft as the base slab was deliberately designed not to cater for the full uplift forces. Readings from water standpipes also showed that ground water did not drop more than 2.5 m during further construction works, due to constant water recharging via the recharge wells. Field observations showed that particles wash-out into the shaft due to recharging did not occur.

1.8 Ground Responses

Finite element analysis was modelled using 2D transient analysis. Figure 5 shows the predicted and measured ground surface settlement profiles at various stages of the excavation. It is found that the ground settlement magnitudes and profiles are sensitive towards the ratio of k_h:k_v of the top organic soil layer, where k_h is the horizontal soil permeability and k_v the horizontal. The predicted and measured values show good agreement. Based on Fig. 5, the predicted ground surface settlement profile using k_h:k_v ratios of 1, 3, 5 and 10 consistently fall within the range of measured settlement values, even though ratio of 5 seems to provide relatively better fit to the measured values in general. Similar anisotropy behaviour in organic soil permeability had also been reported by Mesri et al. (1997), Younger et al. (1997), Sobhan et al. (2007). This could be due to the greater effect of horizontal permeability than vertical within the soft silt layer with organic materials. Therefore, this implies that for any deep excavation works carried out in such soil condition, water seepage can be considerable in the horizontal direction.

2 Case Study 2

Case Study 2 involves an underpass infrastructure to be constructed in the local paludal deposits that comprised organic soils up to 5 m thick. As the underpass design involves formation of side slopes in the peaty soils, some form of ground improvement to stabilise the peaty soil layer is necessary. Therefore, this case study describes the potential effects of anisotropy in permeability of peaty soils and their effects on the ground improvement design.

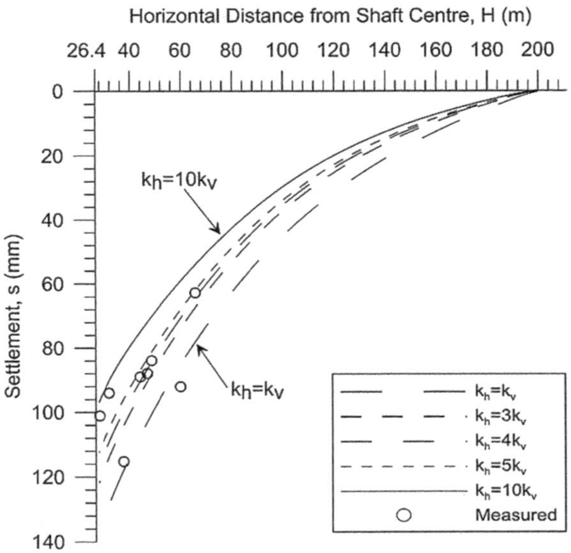

Fig. 5 Predicted and measured ground settlement profiles when excavation depth approaches 13.3 m

Peaty soils with natural moisture content up to 1000% can be found between 1.0 and 4.0 m depths below ground level with the thickness ranging from 0.5 to 5.0 m. Overlying the peaty soil layer is a filled ground. Loose sand and soft clay underlain the peat up to 16.5 m depth. Immediately below this soft clay, is firm to stiff silt or medium sand with thickness ranging from 0.5 to 6.0 m. Very hard silt and dense sand layers are encountered at bottommost soil profile, which consists residual soil of the in situ bedrock.

2.1 Factors Contributing to Ground Settlement

Consolidation settlement occurs when additional surcharge is applied on soft clay and peaty clay, which induces settlement that occurs over a number of years after construction. The magnitude and rate of settlement will vary depending on the thickness of fill and the hydraulic conductivity (permeability) properties of the soft clay and peaty soil which present a time-dependent effect due to relative slow expulsion of pore water pressure, particularly in soft clay.

On the other hand, settlement due to water drawdown has been anticipated because of possible ground water seepage along the slope faces upon excavation, inside the roundabout where the underpasses are constructed. Based on the experience learnt in Case Study 1, where peaty soils can demonstrate serious anisotropy in permeability, finite element modelling was used to estimate the magnitude of ground water drawdown and its impact assessed.

Generally in the settlement analysis, two factors are presumed to be the cause of significant settlement to the ground, namely (i) raising up of the existing road level and (ii) groundwater drawdown due to seepage along the excavated slopes of the underpasses.

2.2 Proposed DSM Treatment of Peaty Soils

After much technical consideration on the appropriate ground improvement technique, 'wet' method of Deep Soil Mixing (DSM) was selected as to enhance the engineering properties of the soft peaty soils. The technique involves the introduction of cement as binder to be mechanically mixed with the existing in situ soft peaty soils. Figure 6 shows the typical section of ground treatment with design parameter c_u of 150 kPa.

The DSM method as proposed in this project is not only to enhance the stability of the slope (due to increased undrained shear strength values) but also to create a physical barrier to limit the ground water drawdown during and after the excavation works due to the relatively permeable peat soil layer. If the DSM barrier is absent, excessive ground water drawdown to as far as 100 m may be triggered with only about 6 m of slope excavation, especially when the underpass alignment is along the peat layer. Therefore, by having the presence of DSM, settlements to nearby existing structures and buildings may thus be minimised. In a parametric study carried out without the presence of DSM, the Factor of Safety (FOS) against slip failure is only 0.775, as shown in Fig. 7.

With the presence of DSM underlining the underpass slopes, a safety factor of 1.258 (FOS $\geq$ 1.2 required) can be achieved, as shown in Fig. 8.

The ground water table of the existing roundabout is at about +5 mRL, and with a 6 m deep excavation (+0.2 mRL), the anticipated water drawdown can be

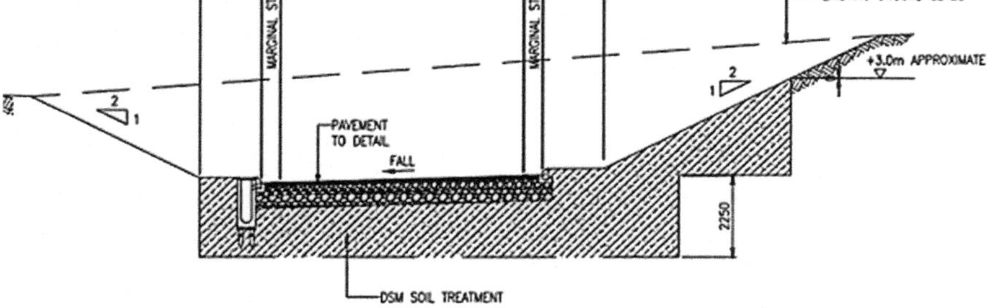

Fig. 6 Typical section of peat layer treated by DSM

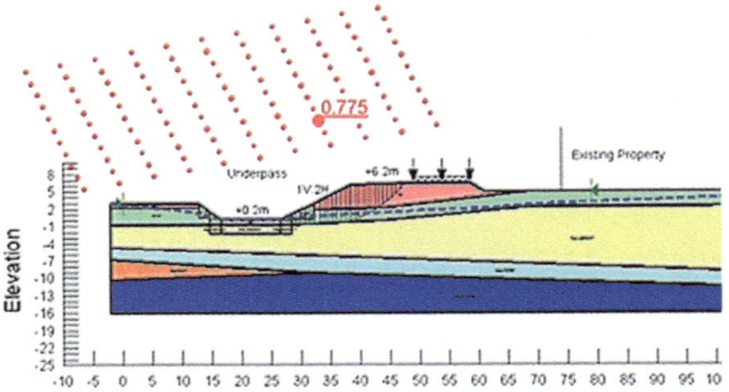

Fig. 7 Computed slope stability without DSM treatment (FOS = 0.775)

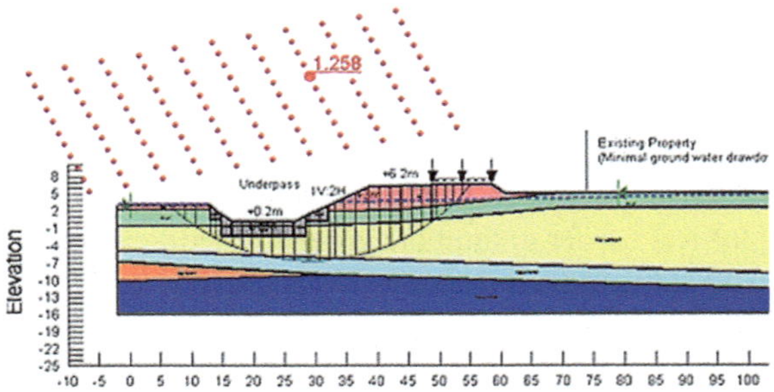

Fig. 8 Computed slope stability with DSM treatment (FOS = 1.258)

substantial in the absence of a ground water barrier. However, with DSM that serves as a ground water barrier, the ground water drawdown may be minimised to +3 mRL.

2.3 *Performance of DSM Installed in Peaty Soils*

Initial design mix comprised approximately about 350 kg/m^3 of cement to be mixed with the in situ peaty soils. The quality of the mixing will then be verified through unconfined compressive strength (UCS) tests and Total Core Recovery (TCR) during the post-installation coring tests. The design criteria for acceptable UCS are minimum 300 kPa and TCR of at least 85%.

From the initial trial post-installation tests, the cored samples from coring points Nos. 1 and 2 (see Fig. 9a, b) seemed not well-formed as most of the samples could not be retrieved. As for coring point No. 3, the sample was retrievable, but failed to meet the design requirement of TCR of more than 85% (see Fig. 9c). Figure 10 shows that 'necking' phenomenon occurred when trial trenching was carried out and the outcomes seemed to tally with both the UCS and TCR results.

In view of the unfavourable results in meeting the design criteria, increased amount of cement dosage was then implemented by the contractor. When cement

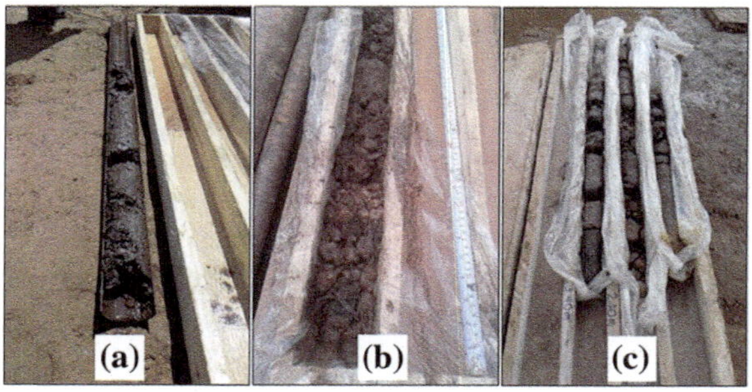

Fig. 9 **a** Non-formation, **b** moderate formation and **c** reasonable formation of DSM treated peaty soils

Fig. 10 Trench excavation of trial DSM piles show 'necking' phenomenon

Fig. 11 Trench excavation shows successful formation of DSM piles in peaty soils

dosage of more than 350 kg/m^3 but less than 500 kg/m^3 was used, the DSM columns in peat formed successfully (see Fig. 11) with the overlapped regions achieving minimum UCS of 300 kPa.

3 Conclusions

The challenges in performing deep excavation in peaty soils have been highlighted in this invited paper and they are not exhaustive. It is imperative and prudent that some form of ground improvement works is necessary to minimise the impact of ground water drawdown during deep excavation works as its impact can be detrimental especially when the anisotropy in permeability of peaty soils, k_h:k_v can range between 2 and 5, with an average of 5 as observed in Case Study 1. In Case Study 2, ground improvement method via wet Deep Soil Mixing (DSM) technique was successfully employed to improve an underlying peaty soil layer with the hope that it could achieve slope stability and to provide an impermeable cut-off prior to carrying out the deep excavation work.

Acknowledgements The author wishes to extend his gratitude to Mohd. Irfaan Peerun for his kind assistance on the editorial work of this paper.

References

Mesri, G., T.D. Stark, M.A. Ajlouni, and C.S. Chen. 1997. Secondary compression of peat with or without surcharging. *Journal of Geotechnical and Geoenvironmental Engineering, ASCE* 123: 411–421.

Ong, D.E.L., and C.S. Choo. 2011. *Sustainable bored pile construction in erratic phyllite*. Kuching: ASEAN-Australian Engineering Congress.

Ong, D.E.L., and C.S. Choo. 2012. Bored pile socket in erratic phyllite of Tuang Formation. In *GEOMATE 2012, 2nd International Conference on Geotechnique, Construction Materials and Environment*, 167–171. KL.

Sobhan, K., H. Ali, K. Riedy, and H. Huynh. 2007. Field and laboratory compressibility characteristics of soft organic soils in Florida. In *Geo-Denver, ASCE, New Peaks in Geotechnics: Advances in Measurement and Modeling of Soil Behaviour*.

Younger, J.S., A.J. Barry, S. Harianti, and R.P. Hardy. 1997. Construction of roads over soft and peaty ground. In *Conference on Recent Advances in Soft Soil Engineering*, Kuching, Sarawak, Malaysia.

Seismic Analysis of Reinforced Soil Retaining Walls

A. Murali Krishna and A. Bhattacharjee

1 Introduction

The reinforced soil walls offer a good solution to conventional earth retaining structures in terms of better utilisation of space, speed of construction and loading capacity. Reinforced soil walls are constructed using different reinforcing elements and wall facing systems. Satisfactory performances and failures of reinforced soil walls during earthquakes are reported by several researchers (Koseki et al. 2006; Koerner and Koerner 2013 etc.). Analysing the performance of retaining structures under static and seismic ground shaking conditions helps to understand better about their behaviour during earthquakes and to design these structures more seismic efficient. Thus, dynamic behaviour of reinforced soil retaining walls is of research interest to several researchers through different modes of studies like, physical model studies, analytical studies and numerical model studies (Cai and Bathurst 1995; Hatami and Bathurst 2000; Ling et al. 2004; Lee et al. 2010; Liu et al. 2011; Krishna and Latha 2012; Bhattacharjee and Krishna 2012, 2015a). This paper highlights the observations obtained from physical land numerical studies on reinforced soil walls subjected to dynamic excitations.

A. Murali Krishna (✉)
Department of Civil Engineering, IIT Guwahati, Guwahati 781032, Assam, India
e-mail: amurali@iitg.ac.in

A. Bhattacharjee
Department of Civil Engineering, Jorhat Engineering College, Jorhat 785007, Assam, India
e-mail: bhatta_arup@yahoo.com

K. Ilamparuthi and R. G. Robinson (eds.), *Geotechnical Design and Practice*,
Developments in Geotechnical Engineering,
https://doi.org/10.1007/978-981-13-0505-4_14

2 Physical Model Studies

Physical model tests are very much essential when the prototype behaviour is very complicated to understand. The use of scaled models in geotechnical engineering offers the advantage of simulating complex systems under controlled conditions, and the opportunity to gain insight into the fundamental mechanisms operating in these systems.

Krishna and Latha (2007, 2009) conducted shaking table tests on wrap-faced- and rigid-faced reinforced soil walls to observe the seismic response. The wall models, tested on the shaking table, were of size 750 mm × 500 mm in plan area and 600 mm (*H*) deep. The models were constructed in flexible laminar container using four layers of geotextile reinforcement of length (L_{rein}) 420 mm (i.e. 0.7*H*) wrapped around to form the facing. The models were constructed in equal lifts of sand filling by pulviation method. For rigid-faced walls, the facing was built from 12 hollow steel box sections and were bolted together with a vertical steel rod. The reinforcements at different vertical spacing were run through the bolts of the facing system to obtain a rigid connection between wall and reinforcements. The model walls were instrumented with displacement transducers, accelerometers and pressures cells. The details of the test configuration and location of various instrumentations (Krishna and Latha 2007) are shown in Figs. 1 and 2.

The models were subjected to sinusoidal motions at different base excitations. Typical response of model, tested for 20 cycles of 0.1 g acceleration (*a*) at 1 Hz frequency (*f*), in terms of horizontal displacements and accelerations at different elevations are shown in Figs. 3 and 4, respectively. The variation displacements of wall facing with frequency, no. of reinforcement layers, surcharge and base acceleration observed by Krishna and Latha (2007) and are shown in Fig. 5. From the figure it is observed that the wall face deformations are higher at low frequency shaking, low surcharge pressure, lesser reinforcing layers and high base acceleration. The model studies were also conducted by varying the relative density of backfill soil. Figure 6 shows the variation of displacements, acceleration amplification and horizontal pressure at different elevations for model with different relative density of backfill soil. The lateral deformation of facing decreases and

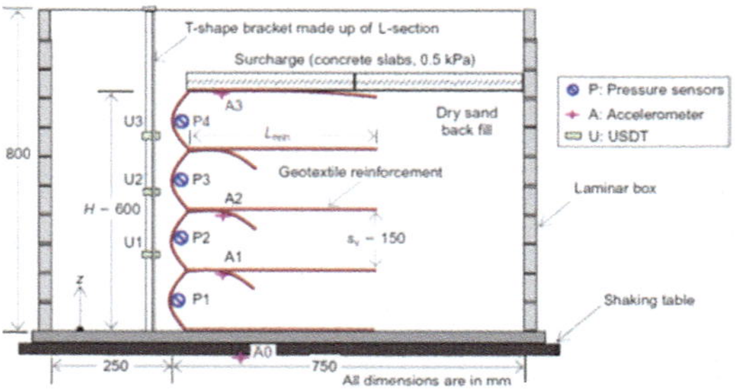

Fig. 1 Schematic diagram of wrap-faced wall configuration (after Krishna and Latha 2007)

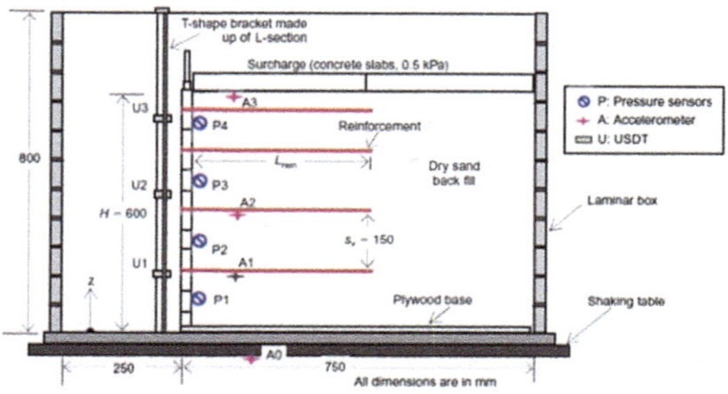

Fig. 2 Schematic diagram of rigid-faced wall configuration (after Krishna and Latha 2009)

acceleration amplifications slightly increases with increase in relative density of backfill soil subjected to higher base excitation.

Krishna and Latha (2009) presented the seismic responses of rigid-faced reinforced soil walls with different reinforcement materials (Fig. 7) such as biaxial geogrid BX1 and BX2, uniaxial geogrid (UA), geonet and weak geotextile (WGT) having ultimate tensile strengths of 26.4, 46.6, 40, 7.6 and 0.4 kN/m respectively.

The inclusion of reinforcing material reduces the horizontal displacement to a considerable extent irrespective of reinforcement stiffness compared with unreinforced wall. More reduction of horizontal displacement for wall reinforced with biaxial geogrid compared to wall with weak geotextile. There is no significant variation in acceleration amplification for reinforced wall with different reinforcement stiffness.

Latha and Krishna (2008) compared horizontal displacements (Fig. 8) and acceleration responses (Fig. 9) of wrap-faced-, rigid-faced unreinforced and rigid-faced reinforced soil walls backfilled with sandy soil at different relative density (RD). The displacements reduce with increase in RD irrespective of facing type and reinforcement. The lateral deformation of wrap-faced wall is more than that of rigid-faced wall. During seismic excitation the soil within geotextile

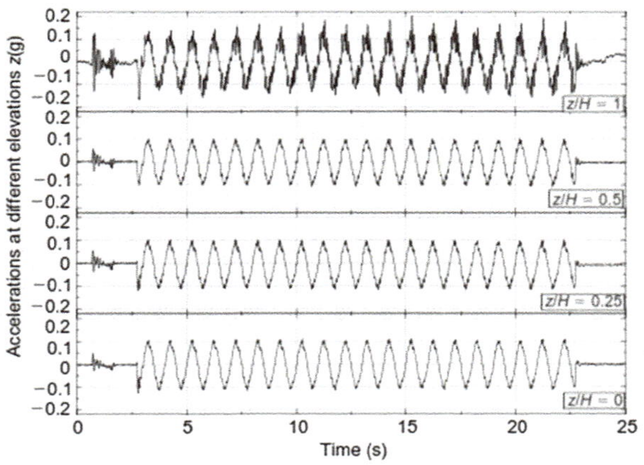

Fig. 3 Accelerations at different elevations (after Krishna and Latha 2007)

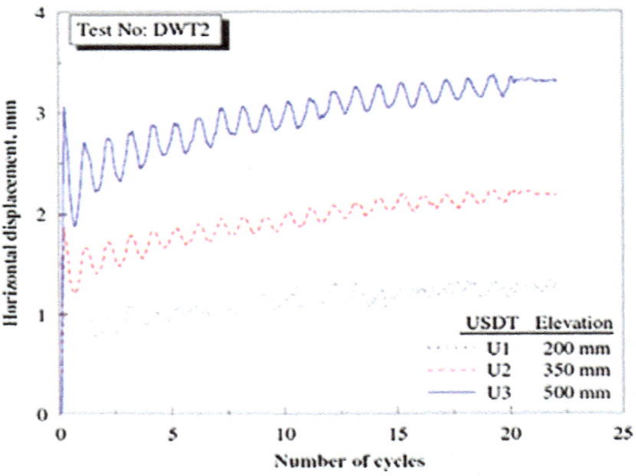

Fig. 4 Typical variation of horizontal displacements (after Latha and Krishna 2008)

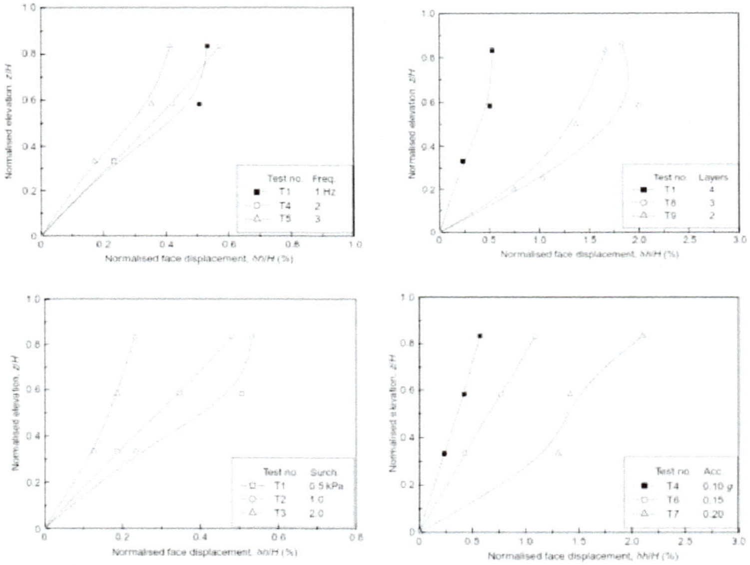

Fig. 5 Variation of horizontal displacement with **a** frequency **b** reinforcement layers **c** surcharge **d** base acceleration (after Krishna and Latha 2007)

wrapped layer settles, as a result face bulges out. But this phenomenon is absent in case of rigid-faced walls. The accelerations are amplified more on top of wall for all three types of wall. But there is no consistent trend in acceleration amplifications with change in relative density of backfill soil in all three types of walls.

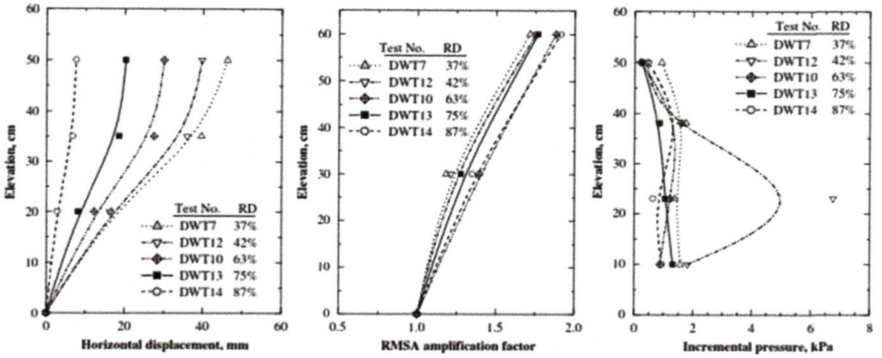

Fig. 6 Response of wrap-faced walls against higher base excitation (at end of 20 cycles of dynamic motion) **a** horizontal displacement **b** acceleration amplification and **c** incremental pressure (after Latha and Krishna 2008)

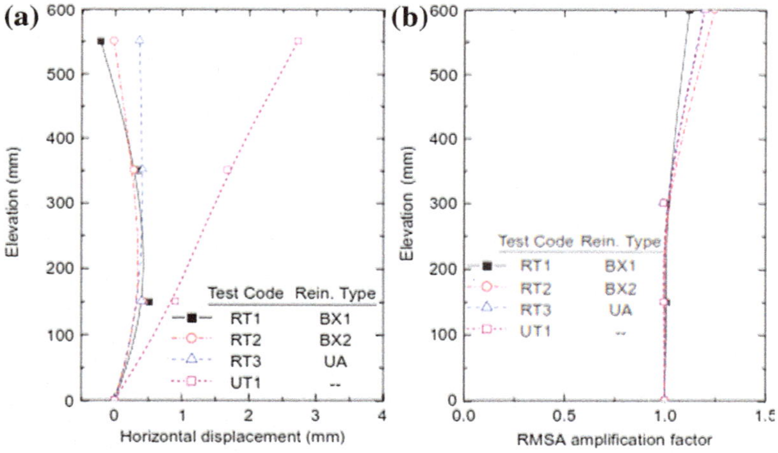

Fig. 7 Response of rigid-faced walls with different types of reinforcement after 20 cycles of 0.1 g at 2 Hz dynamic motion: **a** horizontal displacement **b** acceleration amplification (after Krishna and Latha 2009)

The maximum lateral displacements and maximum acceleration amplification for wrap-faced wall with relative density at smaller excitation (0.1 g, 1 Hz) and higher excitation (0.2 g, 3 Hz) is shown in Fig. 10. The variations of horizontal displacements are more for subjected to higher frequency than that of lower frequency. Small increases in acceleration amplification for wall with denser soil at higher excitation are observed.

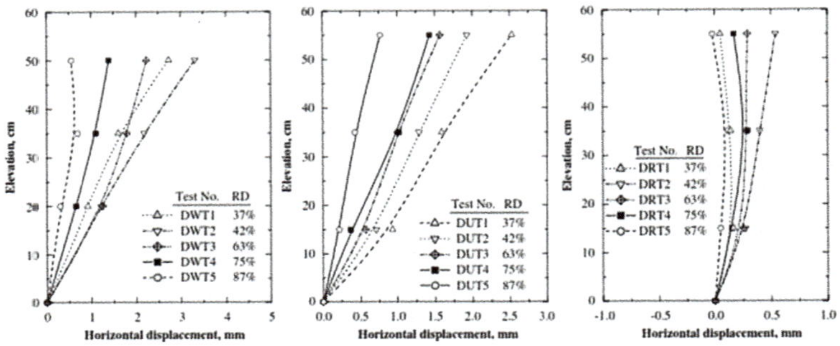

Fig. 8 Displacement profiles after 20 cycles of dynamic excitation for **a** wrap-faced **b** rigid-faced unreinforced **c** rigid-faced reinforced walls (after Latha and Krishna 2008)

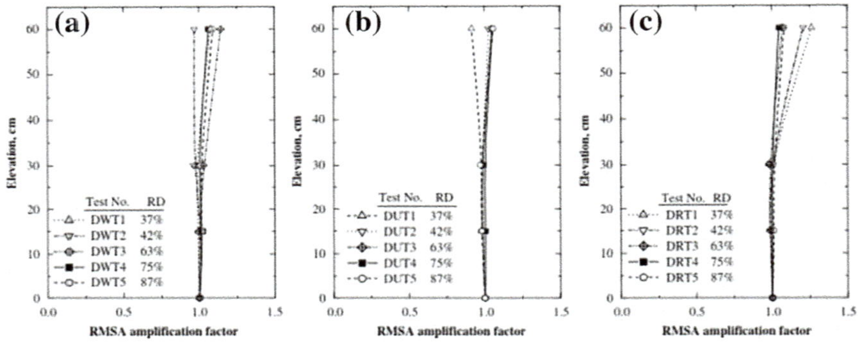

Fig. 9 Acceleration amplification after 20 cycles of sinusoidal dynamic excitation **a** wrap-faced **b** rigid-faced (after Latha and Krishna 2008)

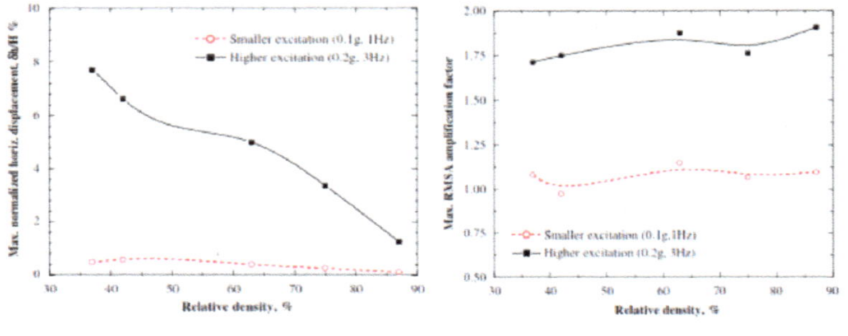

Fig. 10 Variation of **a** maximum displacement **b** maximum acceleration amplification with relative density for wrap-faced wall (after Latha and Krishna 2008)

3 Numerical Model Studies

Numerical models are particularly advantageous because of the difficulties associated with situations in which the prototype structures are too big to be tested; problems related to scaling, instrumentations; and, especially, repetition of model construction, etc. However, the key point to confirm the applicability of any numerical model is by its validation with the available prototype studies and/or small-scale laboratory model studies. The calibrated numerical model can then be used for extensive parametric studies. Krishna and Latha (2012) and Bhattacharjee and Krishna (2012, 2015a) developed numerical models and validated with the physical model tests results. The numerical model of wrap-faced wall was developed by using FLAC^{3D} and is shown in Fig. 11. The validated numerical models were further used to analyse the seismic performance of 6 m high prototype walls. The octahedral shear strains, horizontal and vertical displacements determined along the length of the wall and results are presented in Fig. 12. By comparing strain and displacements, it can be seen that the deformation of wrap-faced wall subjected to seismic excitation consists of three different modes: shear deformation zone within reinforced block, a zone of relative compaction at the end of reinforcement and a shear zone called compound deformation zone extending to the unreinforced backfill.

Bhattacharjee and Krishna (2015b) studied the effect of length of reinforcement on deformation behaviour. Figure 13 shows the comparison of octahedral shear strain in backfill soil with different reinforcement lengths after 20 cycles of dynamic excitation. The octahedral shear strains in soil decrease with increase in reinforcement lengths. The compound deformation zone length decreases with increase in reinforcement lengths. Figure 14 shows the comparison of octahedral shear strain in backfill soil with different number of reinforcing layers after 20 cycles of dynamic excitation. The increase in number of reinforcing layers reduce the soil strain within reinforced zone but do not effect length of compound deformation

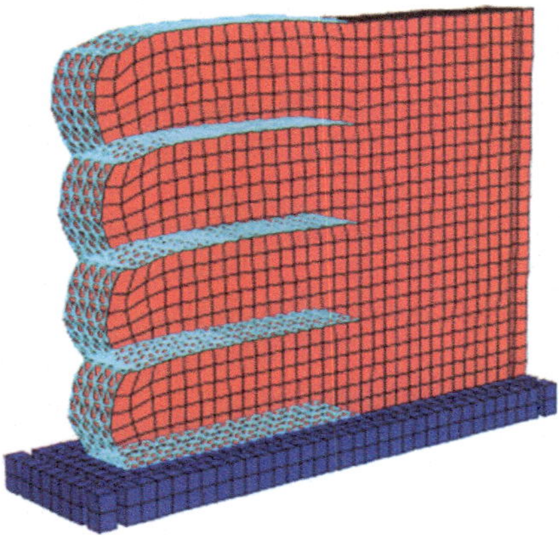

Fig. 11 Numerical model of wrap-faced reinforced soil wall (after Bhattacharjee and Krishna 2012)

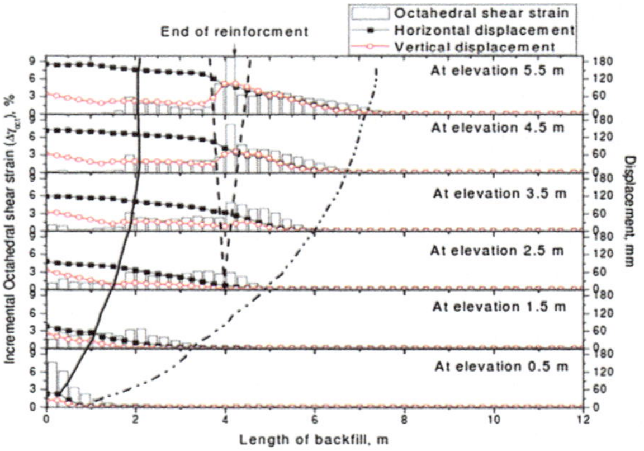

Fig. 12 Octahedral shear strain, horizontal and vertical displacement along the length of backfill (a = 0.2 g, f = 5 Hz) (after Bhattacharjee and Krishna 2012)

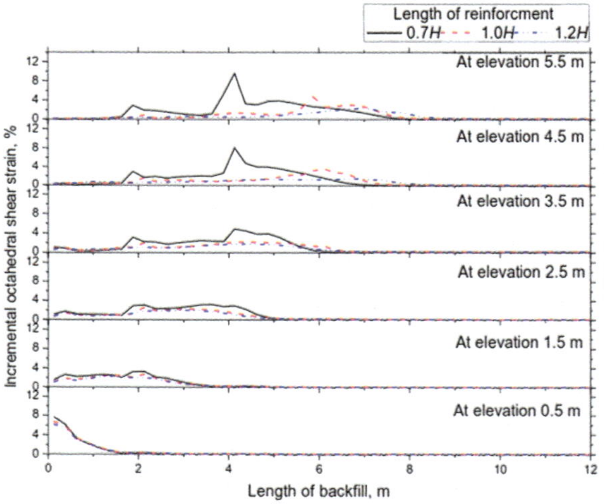

Fig. 13 Comparison of octahedral shear strain at backfill of wrap-faced wall with different reinforcement lengths after 20 cycles of dynamic excitation (a = 0.2 g, f = 5 Hz) (after Bhattacharjee and Krishna 2015b)

zone. So increase in number layers of reinforcements gives more stiffness to the reinforced soil. The study also reported that decrease in friction angles of backfill soil result in extension of the length of compound deformation zone deeper into unreinforced backfill soil.

The numerical model of rigid-faced wall developed by using FLAC^{3D} is shown in Fig. 15. The octahedral shear strains and displacements along the length of backfill between two layers of reinforcements are presented in Fig. 16. By comparing octahedral shear strain, horizontal and vertical displacements two deformation zones are identified. The first zone exists very close to the facing which can be considered as high strain zone and shows relative settlement near wall facing. The second zone is constant strain zone which extends beyond reinforced zone,

Fig. 14 Octahedral shear strains at backfill of wrap-faced walls with different reinforcing layers ($a = 0.2$ g, $f = 5$ Hz) (after Bhattacharjee and Krishna 2015b)

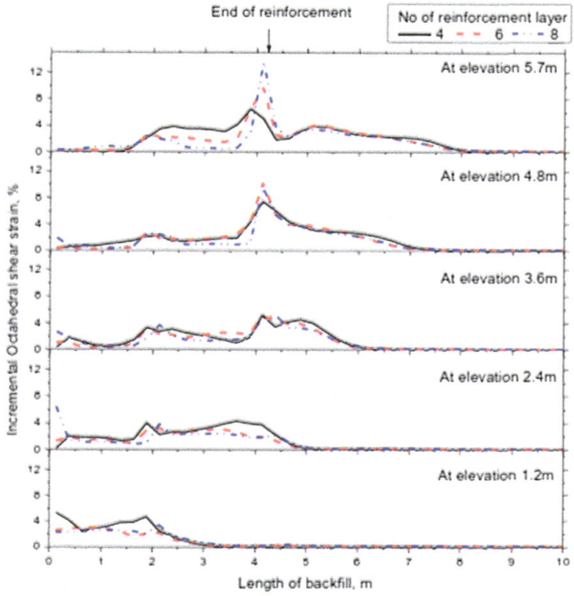

Fig. 15 Numerical model of rigid-faced reinforced soil wall (after Bhattacharjee and Krishna 2015a)

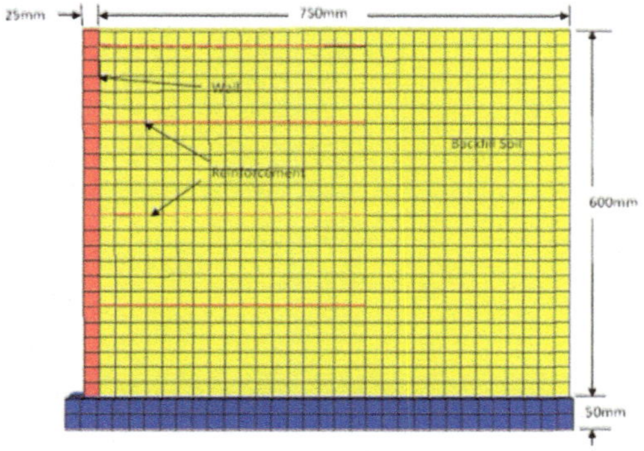

formed due to shear deformation within reinforced zone. This may result in some settlement near the facing and tension cracks in backfill soil.

Krishna and Bhattacharjee (2016) studied seismic behaviour of rigid-faced reinforced soil wall subjected to scaled earthquake ground motion. A full scale calibrated numerical model subjected to five-scaled earthquake ground motions with different predominant frequency ranging from 0.637 Hz for Loma Prieta EQ to 5.437 Hz for Parkfield EQ.

Figure 17 shows the base input ground motions for Loma Prieta and Parkfield EQ and their responses at top. The figure shows that amplitudes close to the fundamental frequency of the wall are amplified the most. Figure 18 shows the

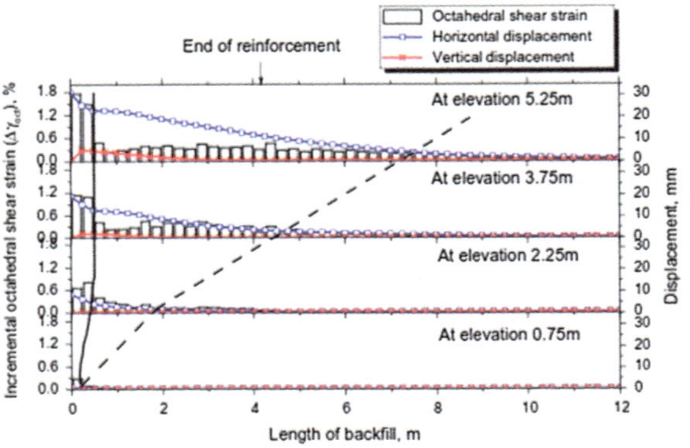

Fig. 16 Octahedral shear strain, horizontal and vertical displacements along length of backfill after 20 cycles of dynamic excitation (a = 0.2 g, f = 5 Hz) (Bhattacharjee and Krishna 2015a, b)

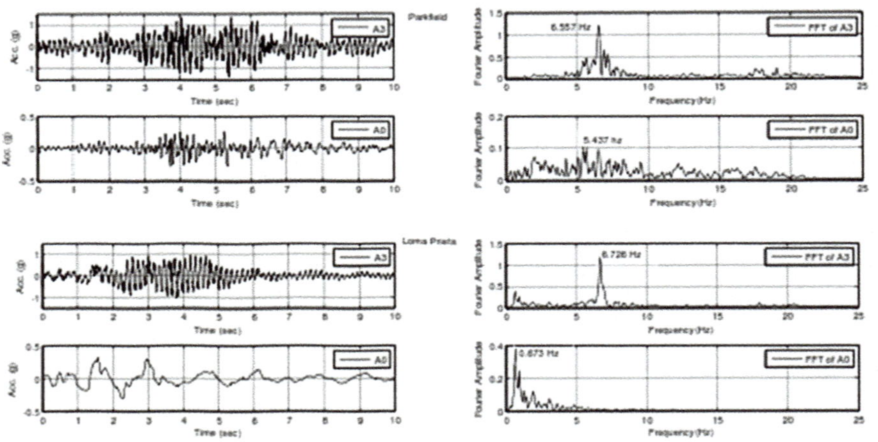

Fig. 17 Acceleration applied at the base of the model (A0) and acceleration recorded at the top of the backfill (A3) and corresponding FFT for Loma Prieta and Parkfield EQ (after Krishna and Bhattacharjee 2016)

variation in the form of horizontal displacement, acceleration amplification and horizontal pressure along the height of wall after different seismic excitations. It is observed from the figure that horizontal displacement and acceleration amplification are different for different earthquake excitations, but the horizontal pressures are nearly identical.

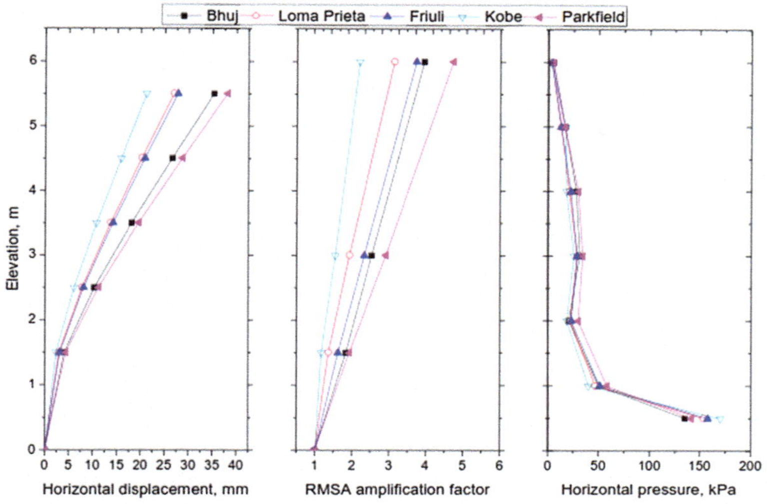

Fig. 18 Horizontal displacements, acceleration amplifications and horizontal pressures for wall subjected to different earthquake excitations (after Krishna and Bhattacharjee 2016)

Fig. 19 Octahedral shear strain along the length of backfill of rigid-faced wall subjected to Bhuj, Kobe and Parkfiled EQ

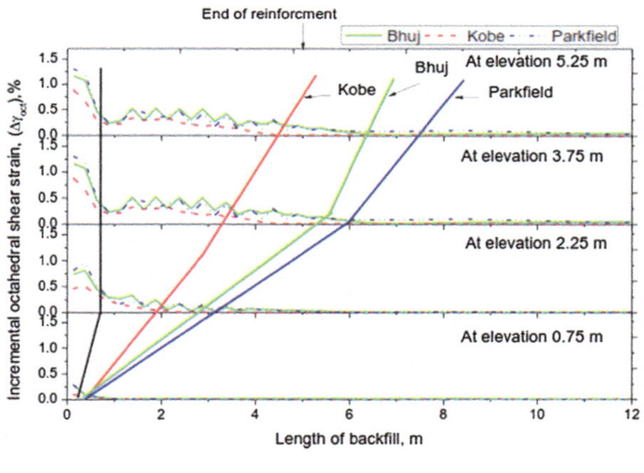

Figure 19 shows octahedral shear strain along length of backfill at different elevations subjected to Bhuj, Kobe and Parkfiled EQ. Two strained zones—a high strain zone near the facing and constant strain zone extended into the backfill are observed. The extent of high strained zone is same for all earthquakes but the extent three constant strained zones is different based on frequency content of scaled earthquakes.

4 Concluding Remarks

The paper discussed about the seismic analysis of wrap-faced- and rigid-faced reinforced soil retaining walls using physical and numerical model studies. Various parameters like backfill RD, reinforcement stiffness, number of layers, length of reinforcement, type of facing influence the wall performance in terms of horizontal displacements, acceleration amplifications, pressures and strains developed in soil and reinforcement. The formations of deformations zones within the wrap-faced- and rigid-faced walls are presented.

With the increasing use of reinforced soil retaining structures in the public infrastructure facilities in large extent; their seismic behaviour must be ensured which can be ascertained through different mode of studies. The insight obtained from the seismic analyses shall be incorporated in the design and construction of such important public infrastructures.

References

Bhattacharjee, A., and A.M. Krishna. 2012. Development of numerical model of wrap faced walls subjected to seismic excitation. *Geosynthetics International* 19 (5): 354–369. https://doi.org/10.1680/gein.12.00022.

Bhattacharjee, A., and A.M. Krishna. 2015a. Strain behavior of backfill soil in rigid faced reinforced soil walls subjected to seismic excitation. *International Journal of Geosynthetics and Ground Engineering*. https://doi.org/10.1007/s40891-015-0016-4.

Bhattacharjee, A., and A.M. Krishna. 2015b. Strain behavior of soil and reinforcement in wrap faced reinforced soil walls subjected to seismic excitation. *Indian Geotechnical Journal* 45 (3): 318–331. https://doi.org/10.1007/s40098-014-0139-x.

Cai, Z., and R.J. Bathurst. 1995. Seismic response analysis of geosynthetic reinforced soil segmental retaining walls by finite element method. *Computers and Geotechnics* 17: 523–546.

Hatami, K., and R.J. Bathurst. 2000. Effect of structural design on fundamental frequency of reinforced-soil retaining walls. *Soil Dynamics and earthquake Engineering* 19 (2000): 137–157.

Koerner, R.M., and G.R. Koerner. 2013. A database, statistics and recommendations regarding 171 failed geosynthetic reinforced mechanically stabilized earth (MSE) walls. *Geotextiles and Geomembranes* 40: 20–27.

Koseki, J., R.J. Bathurst, E. Güler, J. Kuwano, and M. Maugeri. 2006. Seismic stability of reinforced soil walls. In *Proceedings 8th International Conference on Geosynthetics*, 51–78.

Krishna, A.M., and G.M. Latha. 2007. Seismic response of warp-faced reinforced soil-retaining wall models using shaking table tests. *Geosynthetic International* 14 (6): 355–364.

Krishna, A.M., and G.M. Latha. 2009. Seismic behavior of rigid-faced reinforced soil retaining wall models: reinforcement effect. *Geosynthetic International* 16 (5): 364–371.

Krishna, A.M., and G.M. Latha. 2012. Modeling of dynamic response of wrap faced reinforced soil retaining wall. *International Journal of Geomechanics, ASCE* 12 (4): 437–450.

Krishna, A.M., and A. Bhattacharjee. 2016. Behavior of rigid faced reinforced soil retaining walls subjected to different earthquake ground motions. *ASCE International Journal of Geomechanics*. https://doi.org/10.1061/(asce)gm.1943-5622.0000668,06016007.

Latha, G.M., and A.M. Krishna. 2008. Seismic response of reinforced soil retaining wall models: Influence of backfill relative density. *Geotextiles and Geomembranes* 26 (4): 335–349.

Lee, K.Z.Z., N.Y. Chang, and H.Y. Ho. 2010. Numerical simulation of geosynthetic-reinforced soil walls under seismic shaking. *Geotextile and Geomembranes* 28: 317–334.

Ling, H.I., H. Liu, V.N. Kaliakin, and D. Leshchinsky. 2004. Analyzing dynamic behavior of geosynthetic-reinforced soil retaining walls. *Journal of Geotechnical and Geoenvironmental Engineering, ASCE* 130 (8): 911–920.

Liu, H., X. Wang, and E. Song. 2011. Reinforcement load and deformation mode of geosynthetics-reinforced soil walls subject to seismic loading during service life. *Geotextiles and Geomembranes* 29: 1–16.

Tailoring the Properties of Cement-Treated Clayey Soils

S. C. Chian

1 Introduction

With the rapid development of cities worldwide, there are pressures to create space to support continual economic and societal growth. More infrastructures are now constructed underground to free up ground spaces. These infrastructures include basements, utility, train and highway tunnels. Huge amount of unwanted excavated soils are generated as a result. At the same time, the supply of suitable fill materials for land reclamation is becoming scarce in some countries. It would therefore be sustainable to reuse the excavated soils as land reclamation fill materials so as to reduce waste while concurrently create new land for superstructure development. However, in several coastal cities in Asia, such as Bangkok, Shanghai and Singapore, much of the excavated soils are soft clayey materials which are discouraged for use as land reclamation fill materials due to potential excessive settlement and long consolidation time. However, if these excavated soils are improved, the potential use of these unwanted materials can be vast.

Cement stabilisation is a well-established technique commonly adopted for onshore ground improvement in many parts of the world. Large quantities of cement exceeding 25% of the mass of dry soil are usually added to increase the shear strength and reduce compressibility of soft soils. In many cases, ground improvement contracts specify only the minimum unconfined compressive strength performance of cored samples of cement-treated soils. Hence, contractors resort to applying huge amount of cement to ensure that the strength is sufficient to satisfy the requirement.

In the case of land reclamation applications, only a low amount of cement dosage is required to achieve an unconfined compressive strength requirement of

S. C. Chian (✉)
Department of Civil and Environmental Engineering, National University of Singapore, 117576 Singapore, Singapore
e-mail: sc.chian@nus.edu.sg

K. Ilamparuthi and R. G. Robinson (eds.), *Geotechnical Design and Practice*, Developments in Geotechnical Engineering,
https://doi.org/10.1007/978-981-13-0505-4_15

about 280 kPa for the perimeter bunds at the end of 28 days of curing (Lu et al. 2011). However, applying a constant dosage of cement across the whole batch of heterogeneous excavated soil can result in large variation in improved strength. Building foundations will vary in depth in the short term. In the long term, the reclaimed land may suffer uneven settlement. Furthermore, overuse of cement also increases carbon footprint and cost wastage, both which are undesirable to the environment and contractor. A thorough understanding of the strength development is therefore vital to tailor the treated soil to the target strength with an optimal cement dosage. Unfortunately, literature on treated clayey soils with low cement content is limited which motivates the purpose of this research.

This study aims to address the knowledge gap and provide quick estimates of the strength of treated clayey soils. This would enable optimisation of cement dosage to achieve a desirable strength at a particular age of curing.

2 Factors Affecting Strength of Cement-Treated Soils

The strength development of cement-treated soils is chiefly influenced by the properties of the reactants, reaction conditions and sample preparation (Terashi and Kitazume 1999). In this study, different type of soil, cement, mix proportions and curing time are considered to provide some guides on the application of cement-treated soil for land reclamation purposes.

Taki and Yang (1991) carried out unconfined compressive strength test of cement-treated gravel, sand, silt and clay. Their results showed that coarse grained soils produce higher strength than finer grain soil specimens with similar cement content. This is supported by Bell (1993) who found that treated soils with higher content of clay require more quantity of stabilising agent to attain a similar strength.

The type of cement has a significant influence of the development of strength over time. Verástegui Flores et al. (2010) observed that the strength development of Kaolin clay with Portland blast furnace cement is slower in the early stage than that those of ordinary Portland cement due to the presence of slag which retards the hydration process. However, the long-term strength of Portland blast furnace cement can be significantly higher, which is in line with findings of concrete researchers (Roy and Idron 1982; Escalante et al. 2001).

The mix proportions of soil, cement and water also affects the strength of cement-treated soils, with Miura et al. (2001) and Horpibulsuk et al. (2003) identifying the water/cement (*w*/*c*) ratio as the main parameter governing the strength behaviour of cement-treated Bangkok clay. Higher *w*/*c* ratio leads to a rapid reduction in unconfined compressive strength. In the case of soil/cement (*s*/*c*) ratio, at least three zones of strength improvement were identified with cement content (Bergado et al. 1996; Uddin et al. 1997; Zhang et al. 2013). When the cement dosage is very low, the cement is unable to bind the soil together, hence the strength improvement is marginal, as identified as the “inactive zone”. With higher cement dosage, the strength increases significantly and undergoes an “active zone” with large quantity of binding cementitious products generated from hydration and pozzolanic reactions. However, when the cement dosage becomes too excessive that the available amount of water in the mix is insufficient to support the hydration process, the strength development is hindered. Such case is indicated as "inert zone" in the strength versus cement content charts.

Similar to concrete, a longer curation time would permit more complete hydration and pozzolanic reactions to take place. Hence, the strength of the mix would undoubtedly be higher with more extensive and hardened cementitious matrix. A linear relationship of normalised unconfined compressive strength with the logarithm of curing time has been widely reported (Nagaraj et al. 1996; Horpibulsuk et al. 2003; Chian et al. 2015).

3 Experimental Setup

3.1 Materials

The soil used in this study is the Singapore upper marine clay. According to the Unified Soil Classification System (USCS), the marine clay is classified as high plasticity clay (CH). Table 1 summarises the basic physical properties of the clays

In order to better represent field conditions where sand impurities are present, fine grained sand is added into Singapore marine clay to study the influence of coarse grain particles in clay. The admixed sand has a specific gravity of 2.63 and bulk density of 1500 kg/m^3. As shown in the soil grading curve in Fig. 1, the

Table 1 Basic properties of Singapore upper marine clay and Kaolin clay

Properties	Singapore marine clay	Kaolin clay
Liquid limit (%)	70–90	60.5
Plastic limit (%)	36–56	42.5
Specific gravity	2.62–2.69	2.60

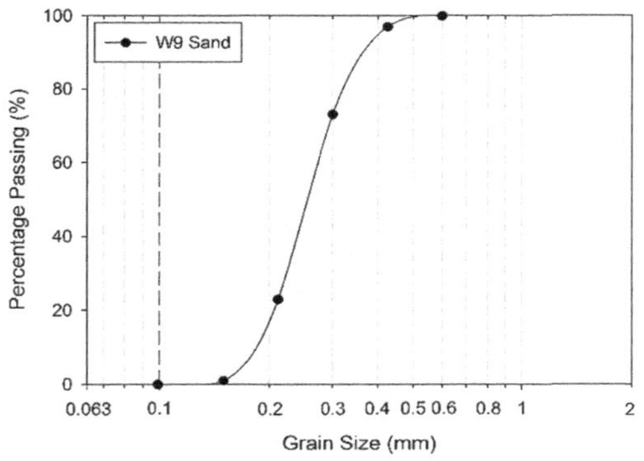

Fig. 1 Particle size distribution of W9 sand

Table 2 Basic properties of Portland blast furnace cement (PBFC)

Physical Properties	Value
Density	2900 kg/m^3
Fineness	429 m^2/kg
Initial setting time	125 min
Final setting time	225 min
Soundness	<0.5 mm
Consistency	29.6%
Compressive strength (28-day)	40 MPa

particle size distribution is poorly graded, with a mean effective diameter (D_{50}) of 0.25 mm. The uniformity coefficient (C_u) and curvature coefficient (C_c) are 1.43 and 1.07 respectively.

Portland blast furnace cement (PBFC) with 85% of slag content, supplied by Engro Corporation Ltd Singapore, was used as the stabilising agent in this study. Its lower early strength gain is beneficial for early quality control. The basic properties of PBFC are presented in Table 2.

3.2 Preparation

The raw Singapore Marine Clay was sieved for removal of stones and shell pieces. The sand fraction was less than 5% and hence, its influence is deemed negligible. The amount of PBFC and water to be added were calculated based on the total soil weight (clay plus sand, if any). For mixes involving the study of sand impurities, the desired sand content was calculated based on the total dry weight of soil.

The Japanese standard JGS 0821 (Japanese Geotechnical Society 2000) was adhered with the mixing process maintained at 10 min in total, of which 1 min was allocated for manual mixing and scraping of materials attached to the sidewall and bottom of the mixing bowl. After mixing, the mixture was transferred to cylindrical split moulds of 50 mm in diameter and 100 mm in height. The specimen was compacted in three layers by manual tamping to minimise entrapped air voids. Next, the specimens were fully immersed in water and stored under constant temperature for curing (23±2 °C). Prior to testing, the top and bottom of the specimen were trimmed flat to 100±5 mm to maintain a length to diameter ratio of about 2.

During mixing, it was observed that the presence of sand led to a slurry mixture. The higher the percentage of sand content, the more slurry was the mixture. This is expected as sand does not retain water as significantly as clay. Liquid limit (w_L) of the soil mixtures was therefore determined using the cone penetration test according to BS 1377-2 (1990) to assess their water holding capacity. Liquid limit is essentially a measure of lower strength limit of shearing resistance, representing the water content at which soil approaches liquid state. Skempton (1953), Seed et al.

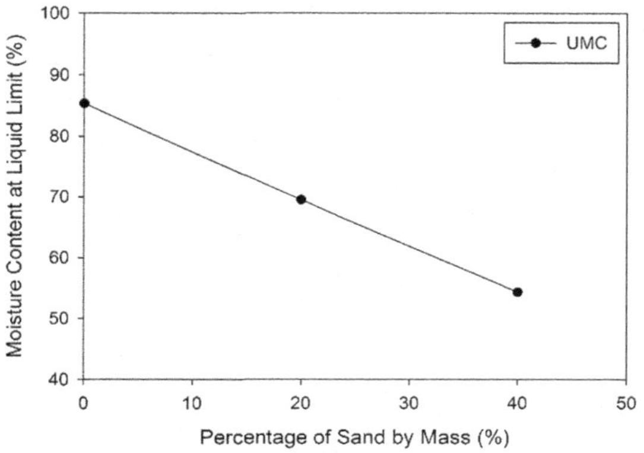

Fig. 2 Liquid limit of Singapore marine clay

(1964), Pandian and Nagaraj (1990) indicated that the composite effects of soil constituents and their interactions with pore fluid are largely reflected in their liquid limit (w_L) (Fig. 2).

3.3 Testing

Since the focus of this study was for land reclamation purposes, the mix proportions were made representative of those used in the field. The cement content used in land reclamation is generally within 10–18% of the dry mass of soil. The range of water/cement ratio considered in this study was between 8.5 and 13. The curing duration was between 3 days to 91 days. For specimens used to investigate the effects of sand impurities, the sand content as a percentage of the total soil were 0, 20 and 40%.

Unconfined compressive tests following BS 1377-7 (1990) were carried out in this study. Bender element tests were also performed to obtain estimates of the small-strain shear modulus of the treated specimens.

4 Strength Development of Cement-Treated Clayey Soils

4.1 Effect of Mix Proportions and Curing Time

Figure 3 shows the influence of water/cement and soil/cement ratios on the unconfined compressive strength of Singapore marine clay at different curing time. It was observed that lower water/cement ratio produces a higher strength of the specimen. This is analogous to concrete technology. In the case of cement-treated soils, the soil/cement ratio is often higher in value and should be studied to assess its significance on the strength development.

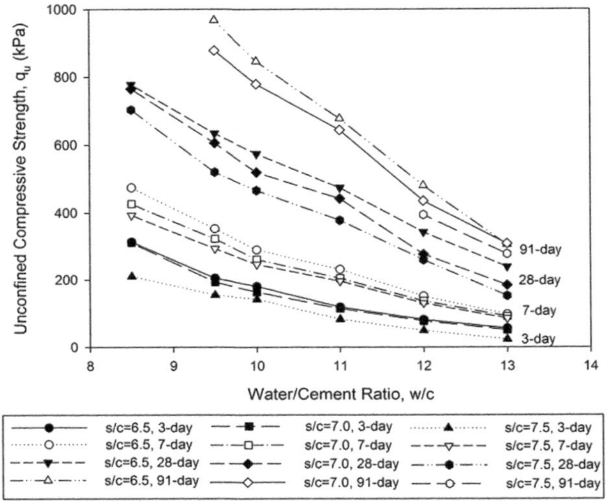

Fig. 3 Effect of mixing proportions on unconfined compressive strength of Singapore marine clay

Fig. 4 Effect of curing time on normalised unconfined compressive strength

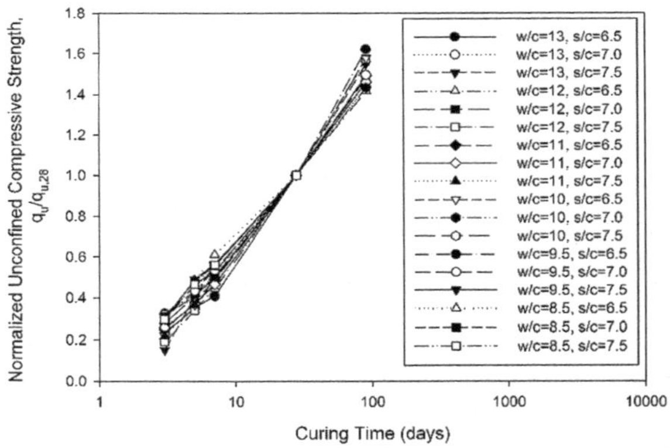

In Fig. 3, the increase in soil/cement ratio leads to a reduction in unconfined compressive strength of the specimen. This is logical as a higher soil/cement ratio would imply a relatively lower amount of cement to soil. Hence, there is more cement available to bind the soil particles together, thereby producing a stiffer material.

The curing time also have a significant effect on the strength of cement-treated specimens. With longer curing time, more thorough hydration reaction and pozzolanic reactions can be achieved, thereby achieving a stronger specimen.

In order to observe the gain in strength of cement-treated Singapore marine clay with respect to the conventional 28-day strength requirement, a normalised strength versus curing time is plotted in Fig. 4. In the figure, the points fall along a near

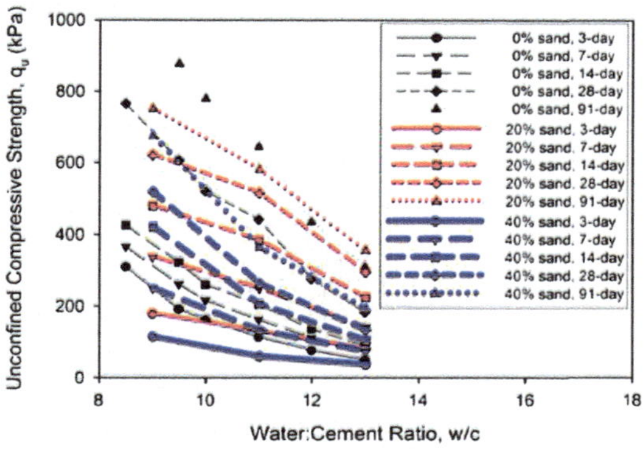

Fig. 5 Effect of sand impurities on unconfined compressive strength of Singapore marine clay

linear relationship in the semi-log scale, indicating a convenient logarithmic trend for the strength development with time. It is also noted that the same trend persists for specimens over a range of *w*/*c* and *s*/*c*, implying that the development characteristics of different mix proportions are comparable.

4.2 *Effect of Sand Impurities*

The effects of sand impurities in Singapore marine clay is illustrated in Fig. 5. At the same soil/cement ratio, it can be observed that the reduction in unconfined compressive strength with increasing water/cement ratio persists for clays with sand impurities up to 40% of the total soil dry mass. It was also observed that the increase in sand content resulted in a lower strength as shown in the figure. This is appears in conflict with the findings by Taki and Yang (1991), who showed higher strength with cement content for coarser grained cement-treated soils. However, the cement content proposed by the literature did not consider the water content in the soil, therefore producing differing conclusions.

With the same mass of soil, a higher percentage of sand would lead to a lower liquid limit which infers that lesser amount of water is absorbed by the soil. Therefore, more "free" water is available to the cement, thereby reducing the strength of the mix as reflected in Fig. 5.

5 Proposed Empirical Strength Model

Having identified water/cement ratio as a major factor affecting the unconfined compressive strength of cement-treated soils, the Abrams' law, a common cementitious strength model was used to depict the influence of mix proportions on the strength development of cement-treated clay. It was demonstrated earlier in

Fig. 4 that the effect can be illustrated with a semi-log relationship with time. The Abrams' equation can therefore be modified with the addition of a natural logarithmic term. In the case of the effect of soil/cement ratio, it was observed that the change in the ratio leads to a vertical shifting of the curve in Fig. 3. The eventual proposed empirical strength model, encompassing the effects of water/cement (*w/c*), soil/cement (*s/c*) and curing time (*t*), is as shown in Eq. (1).

$$q_{\mathrm{u}} = \frac{a + b\left(\frac{s}{c}\right)}{Y^{w/c}} \ln(t) \tag{1}$$

The parameters *a*, *b* and *Y* are fitting parameters depending on the type of cement and clay.

Figure 6 shows the goodness of fit between the predicted and measured unconfined compressive strength for the full range of tests on pure Singapore marine clay in this study using *a, b* and *Y* values of 3700, −75 and 1.35 respectively. This justifies the robustness of the equation in estimating the strength development of cement-treated Singapore marine clays. Further validation with Bangkok, Ariake, Kaolin, Black and Yangtze River clays with ordinary Portland cement are discussed in Chian et al.'s (2015).

In the case of clay specimens with sand impurities in Fig. 5, there exist no clear trend of strength development between different percentages of sand content. The conventional water/cement ratio is therefore no longer relevant in depicting the strength development of cement-treated clayey soils with sand impurities. In order to incorporate the effect of sand, the "*liquid limit ratio* (w/w_L)" was introduced to describe the state of soil microfabric and amount of free-water involved (Chian et al. 2016). w/w_{L} is defined as the soil's water content divided by the water content at liquid limit.

Figure 7 shows the predicted versus measured unconfined compressive strength of the cement-treated clayey soil with sand impurities, which indicate the suitability of the new liquid limit ratio in replacement of the conventional water/cement ratio parameter in the proposed predictive strength model.

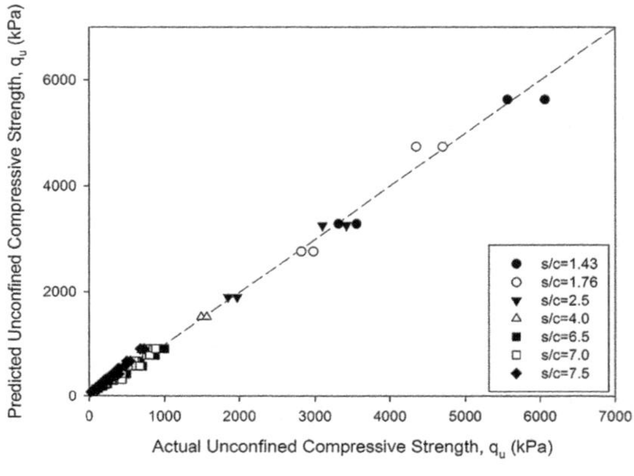

Fig. 6 Predicted versus actual unconfined compressive strength of Singapore marine clay

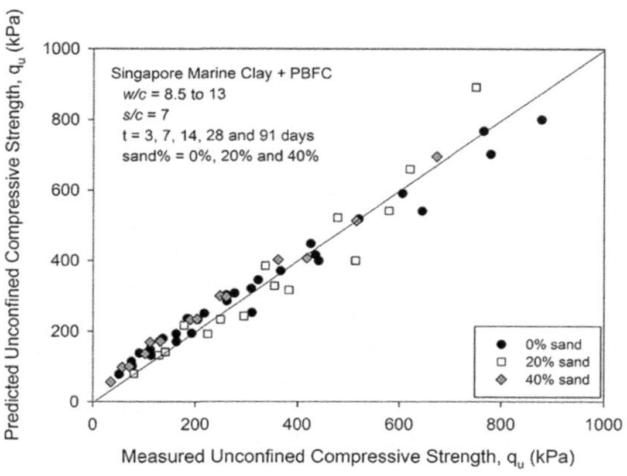

Fig. 7 Predicted and measured strength of cement treated clayey soils with sand impurities

6 Conclusion

A comprehensive study was carried out to investigate the influence of mix proportions, curing time and sand impurities. From the analysis, the following conclusions were drawn:

- The proposed predictive strength model was shown to be capable of producing matching prediction of strength over different clay and cement types, mixing ratios and curing time. Engineers may now easily estimate the strength development of cement-treated clays at different *w*/*c*, *s*/*c* and *t* by conducting limited mix trials based on the proposed strength model in this study.
- The inclusion of sand content yields a noticeable strength reduction in cement-treated clay. This may have significant implication to cement stabilisation practice, where the dosage of cement in ground improvement may have to be altered when encountering sandy-clays in the field to achieve a desired final strength.
- The conventional water/cement ratio (*w*/*c*) has to be altered in the strength model to encompass the effects of lower water-affinity sand grains. The use of a novel '*liquid limit ratio* $(w/w_L)/c$' is more appropriate and produce good estimates of cement-treated clay with sand impurities.

References

Bell, F.G. 1993. *Engineering treatment of soil*. London: Spon.

Bergado, D.T., L.R. Anderson, N. Miura, and A.S. Balasubramaniam. 1996. *Soft ground improvement in lowland and other environments*. New York: ASCE Press.

BS 1377-2. 1990. *Methods of test for soils for civil engineering purposes—Part 2: Classification tests*. London: British Standards Institution.

BS 1377-7. 1990. *Methods of test for soils for civil engineering purposes—Part 7: Shear strength tests (total stresses)*. London: British Standards Institution.

Chian, S.C., S.T. Nguyen, and K.K. Phoon. 2015. An extended strength development model of cement treated clay. *Journal of Geotechnical and Geoenvironmental Engineering,*. https://doi.org/10.1061/(ASCE)GT.1943-5606.0001400,06015014.

Chian, S.C., Y.Q. Chim, and J.W. Wong. 2016. Influence of sand impurities in cement-treated clays. *Geotechnique*. https://doi.org/10.1680/jgeot.15.P.179.

Escalante, J.I., L.Y. Gomez, K.K. Johal, G. Mendoza, H. Mancha, and J. Mendez. 2001. Reactivity of blast furnace slag in Portland cement blends hydrated under different conditions. *Cement and Concrete Research* 31: 1403–1409.

Horpibulsuk, S., N. Miura, and T.S. Nagaraj. 2003. Assessment of strength development in cement-admixed high water content clays with Abrams' law as a basis. *Geotechnique* 53 (4): 439–444.

Japanese Geotechnical Society, JGS. 2000. *Practice for making and curing stabilized soil specimens without compaction*. JGS T 0821-2000, Japanese Geotechnical Society (in Japanese).

Lu, Y., T.S. Tan, and K.K. Phoon. 2011. *Accelerated testing of cement treated Singapore marine clay cured under elevated temperature*. Technical report. National University of Singapore, Centre for Soft Ground Engineering, Singapore.

Miura, N., S. Horpisulbuk, and T.S. Nagaraj. 2001. Engineering behaviour of cement stabilized clay at high water contents. *Soils and Foundation* 41 (5): 33–45.

Nagaraj, T.S., P. Yarigar, N. Miura, and A. Yamadera. 1996. Prediction of strength development cement admixture based on water content. In *Proceedings of the 2nd International Conference on Ground Improvement Geosystems*, vol. 1, 431–436.

Pandian, N.S., and T.S. Nagaraj. 1990. Critical reappraisal of colloidal activity of clays. *Journal of Geotechnical Engineering* 116 (2): 285–296.

Roy, D.M., and G.M. Idorn. 1982. Hydration, structure and properties of blast furnace slag cements, mortars and concrete. *ACI Journal* 79 (6): 444–457.

Seed, H.B., R.J. Woodward, and R. Lundgren. 1964. Fundamental aspects of Atterberg limits. *Journal of Geotechnical Engineering* 117 (9): 1288–1330.

Skempton, A.W. 1953. Colloidal activity of clays. In *Proceedings of the 3rd International Conference on Soil Mechanics and Foundation Engineering,* Zurich, Switzerland, 57–67.

Taki, O., and D. Yang. 1991. Soil-cement mixed wall technique. *Geotechnical Engineering Congress, ASCE, Special Publication* 27: 298–309.

Terashi, M., and M. Kitazume. 1999. QA/QC for deep-mixed grout: current practice and future research needs. *Ground Improvement* 164 (3): 161–177.

Uddin, K., A.S. Balasubramaniam, and D.T. Bergado. 1997. Engineering behavior of cement-treated Bangkok soft clay. *Geotechnical Engineering* 28 (1): 89–119.

Verástegui Flores, R.D., G. Di Emidio, and W. Van Impe. 2010. Small-strain shear modulus and strength increase of cement-treated clay. *Geotechnical Testing Journal* 33 (1): 62–71.

Zhang, R.J., A.M. Santoso, T.S. Tan, and K.K. Phoon. 2013. Strength of high water-content marine clay stabilized by low amount of cement. *Journal of Geotechnical and Geoenvironmental Engineering* 139 (12): 2170–2181.

Landslides in Nilgiris: Causal Factors and Remedial Measures

S. S. Chandrasekaran, V. Senthilkumar and V. B. Maji

1 Introduction

Nilgiris district is located in Western Ghats in Tamil Nadu state in India and surrounded by Coimbatore and Erode districts of Tamil Nadu state on the east, Kerala state on the west and Karnataka state on the north. The district is a hilly region and the highest peak of Doddabetta lies at an elevation of 2595 m from the Mean Sea Level (MSL) (Nilgiris 2011). Numerous landslides have occurred in Nilgiris in the past with the frequency of occurrence increased to alarming levels in last three decades. Some of them created severe damage to infrastructures like road, rail routes and buildings. As the Nilgiris district receives heavy rainfall both in South West and North East monsoons, rainfall is a major triggering factor for landslides in Nilgiris. Landslides occurred at more than three hundred locations of the district in November 2009 alone, which resulted in direct and indirect loss to the government and people (Ganapathy et al. 2010). Landslides in 2009 left 50 people dead and hundreds homeless. Mettupalayam–Ooty National highway (NH 67), the life line of Nilgiris and Mettupalayam–Coonoor rail route of Nilgiris Mountain Railway (NMR) got severely affected at many places (Fig. 1) (Chandrasekaran 2010; Elayaraja et al. 2015; Senthilkumar et al. 2015). The heavy rainfall in July

S. S. Chandrasekaran (✉) · V. Senthilkumar
Geotechnical Engineering, SCALE, VIT University, Vellore, India
e-mail: chandrasekaran.ss@vit.ac.in

V. Senthilkumar
e-mail: senthilkumar.vadivel@vit.ac.in

V. B. Maji
Department of Civil Engineering, IIT Madras, Chennai, India
e-mail: vbmaji@iitm.ac.in

K. Ilamparuthi and R. G. Robinson (eds.), *Geotechnical Design and Practice*,
Developments in Geotechnical Engineering,
https://doi.org/10.1007/978-981-13-0505-4_16

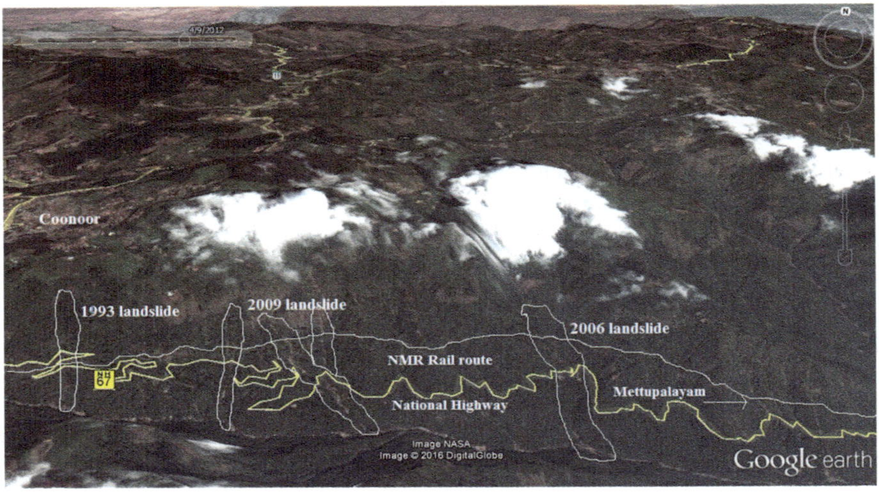

Fig. 1 View of Mettupalayam–Coonoor road and rail network and major landslides occurred

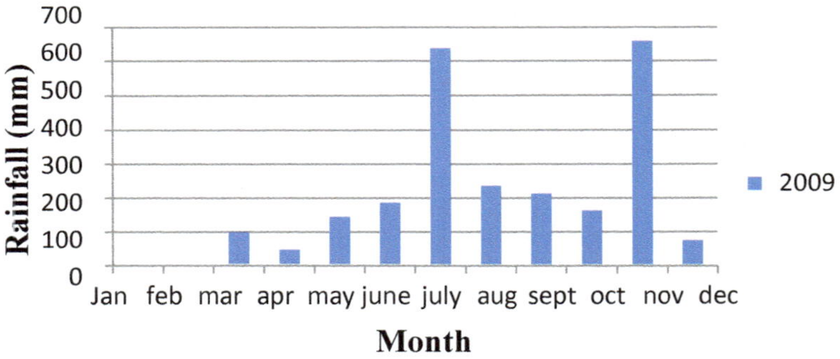

Fig. 2 Nilgiris monthly rainfall during 2009

2009 (during Southeast monsoon) followed by very heavy rainfall in November 2009 (during Northeast monsoon) was the main triggering factor for the 2009 landslides in Nilgiris (Fig. 2) (IMD 2016). In addition to rainfall, there are some causal factors involved in triggering of landslides and are discussed in detail in the present study. Stability of slope can be increased by providing suitable remedial measures. The defects of existing remedial measures practiced in Nilgiris and suggestion of suitable advanced remedial measures have been discussed in this paper.

2 Causal Factors Involved in Triggering of Landslides in Nilgiris

In every slope, there are forces which tend to promote downslope movement and opposing forces which tend to resist movement. The factor of safety of a slope is a ratio of shear strength of the soil to the downslope shear stress, along an assumed or known rupture surface (Popescu 2002). Based on general definition of factor of safety, Terzaghi (1950) divided causes of landslide into two types, external and internal causes. External causes include the changes in geometry of slope, loading at the slope crest, unloading the slope toe, shocks and vibrations, drawdown, changes in water regime. Internal causes include progressive failure, weathering and seepage erosion. External and internal causes may involve in either increasing the shear stress or reducing the shear resistance and some of them can affect both terms of factor of safety simultaneously (Varnes 1978). The significant causal factors of landslides at Nilgiris are discussed below.

2.1 Excavation of Slope at Toe

Figure 3a, b shows the excavation of slope at toe for widening of road along Mettupalayam–Coonoor highway. Excavation of slope at toe reduces the shear strength of the soil due to yielding of soil mass (Sutejo and Gofar 2015). When rainfall infiltration saturates the slope, the shear stress increases thus increases the driving force. Decrease in shearing resistance and increase in shearing stress due to excavation of slope at toe and rainfall infiltration may lead to slope failure. Hence, the slope failure may end up with complete closer of road traffic in Nilgiris.

Fig. 3 View of excavation of slope at toe; **a** and **b** along Mettupalayam–Coonoor road network

Fig. 4 View of vertical cut; **a** all around the building **b** near to building foundation

2.2 Unplanned Vertical Cut

Figure 4a shows the vertical cut made all around the building located near Kattabettu for about 15.00 m height. The lateral resistance of the slope has been completely removed all around the building due to unplanned vertical cut. Removal of lateral resistance will lead to reduction in shear resistance and increase in shear stress thus result in slope failure. A building located at the crest of a slope in Kodappamandu near Ooty has been collapsed due to vertical cut nearer to the foundation of the house (Fig. 4b). Loading at slope crest and vertical cut increase shear stress and reduce the shearing resistance of soil that led to slope failure during heavy rainfall in November 2009.

2.3 Loading the Slope at Crest

Figure 5a, b shows slope failure at Kothagiri and Nondimedu where buildings constructed at slope crest are on the verge of failure. Failure occurred, due to loading of slope at the top, under heavy rainfall. Building weight present on the slope crest adds to gravitational load which increase the driving force and causes slope failure. If overloading occurs away from the crest of the slope the influence is

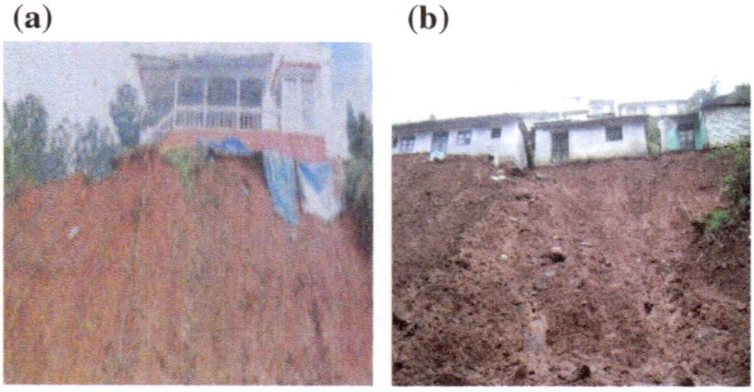

Fig. 5 View of slope failure due to loading the slope at crest; **a** Kothagiri **b** Nondimedu

Fig. 6 View of blocking of surface drainage pipe and drainage culvert; **a** and **b** along Mettupalayam–Coonoor road network **c** Aravankadu location

minimal but if it is within the vicinity of the crest of the slope its effect cannot be ignored as observed from Fig. 5a, b.

2.4 Blocking of Drainage Systems

Drainage of water is one of the most effective techniques which increase stability of slopes. Figure 6a–c shows blocking of surface drainage systems (drainage pipes and drainage culverts) along Metupalaym–Coonoor national highway and Aravankadu location. Blocking of drainage system allows the rain water to infiltrate into the slope, leads to build up of pore water pressure and thereby increases the driving force.

2.5 Vegetation Removal and Dumping of Loose Soil

Figure 7a shows removal of vegetation in a slope located at Chinnabikatty near Coonoor. Generally plant roots provide a strong interlocking network to hold unconsolidated materials together and prevent flow (Coppin and Richards 1990). In addition to that, plants are useful in removing of water from the soil which increases the shear strength. The removal of slope vegetation in Nilgiris may result in increasing either the rate of erosion or the frequency of slope failure. Figure 7b shows the dumping of loose soil on the slope surface near Coonoor. The additional weight due to dumping increases the shear stress. Whenever excavation of slope is made at the toe for widening of road, the unstable loose soil mass on the slope surface slide down and will disturb the road traffic.

Fig. 7 View of causal factors; **a** removal of vegetation on slope surface **b** dumping of loose soil on the slope surface

3 Remedial Measures

The remedial measures implemented so far in Nilgiris are construction of retaining walls and gabion walls (Fig. 8a, b). It was observed that the weep holes provided in the retaining walls often got blocked with fine particles. This could be due to malfunctioning of the filtering arrangement provided in the weep hole during the time construction of retaining walls. The similar problem has been observed in many locations throughout the district.

Surface drainage systems provided throughout the district is also blocked due to lack of maintenance. Excess pore pressure and seepage forces during the heavy rainfall decrease the effective stress of the soil and thus reduce the shear strength, consequently resulting in slope failure (Anderson and Sitar 1995; Cascini et al. 2010; Chandrasekaran et al. 2013a). Hence a proper subsurface drainage system and their periodic maintenance can be an effective solution in stabilization of slopes to avoid the slope failure under the influence of rainfall in Nigiris (Bhandari 2006). India's first subsurface drainage system by horizontal drains is implemented in Porthimund dam in Nilgiris district. The landslide occurred on the downstream side of dam in 1979 due to the excess hydrostatic pressure build up in the soil mass due to the seepage of drain water at the higher reaches of the slope (Natarajan et al. 1984). In order to dissipate the excess hydrostatic pressure from the soil mass overlying the bed rock, the subsurface drainage technique with horizontal drains in combination with vegetative turf has been implemented as appropriate remedial measures to control the landslides (Natarajan et al. 1984). The same technique

Fig. 8 Remedial measures; **a** view of retaining wall **b** view of gabion wall

needs to be implemented in other locations in Nilgiris. In addition to the sub surface drainage system some of the additional techniques including internal slope reinforcement consist of soil nails and stabilizing piles have to be explored. Slope failure due to vertical cut can be controlled by adopting nailing technique. The sliding mass above the failure surface can be strengthened by placing passive piles embedded into sufficient depth in the stable layer below. The loads can be transferred to the more stable underlying layers and the soil movements can be resisted by piles which act as a barrier (Kourkoulis et al. 2011). So far the present paper discussed about different causal factors involved in triggering of landslides in Nilgiris and different remedial measures were suggested to control the effects of landslides in Nilgiris. The following sections will give an over view of numerical analysis pertaining to 2009 landslides in Nilgiris and study on evaluation of potential of earthquake induced landslides in Nilgiris carried out by authors. The on-going research work on detailed site specific geotechnical investigation being carried out by authors is also discussed in the following sections.

4 Numerical Analysis Pertaining to 2009 Landslides in Nilgiris

Rainfall induced landslides have occurred at more than three hundred locations in Nilgiris during north east monsoon in November 2009 resulted in loss of 50 lives and property loss of about three hundred crores (Chandrasekaran et al. 2013b). In order to understand its failure mechanism and effects of rainfall and other causal factors, numerical analysis has been performed by Chandrasekaran et al. (2013b). Different case histories have been identified where the landslides occurred in 2009 which includes slope failure which supported railway track at Aravankadu (Fig. 9a), failure of retaining structure supporting building at Coonoor (Fig. 9b) and slope and retaining wall failure which supported a road network (NH-67) at Chinnabikkaty (Fig. 9c) (Chandrasekaran et al. 2013b). Soil samples have been collected from three landslide locations and laboratory investigations have been carried out. From the laboratory investigations, it is observed that soils have high fines content and low values of hydraulic conductivity. The classification of soil for three locations are Silty sand (SM) for Aravankadu, Lean clay with sand (CL) for Coonoor and Sandy silt (ML) for Chinnabikkaty. A finite element analysis has been carried out using PLAXIS 2D programme for all three locations. Numerical analysis reveals that increase in pore pressure under heavy rainfall reduces the shear strength of the soil and consequently resulted in progressive failure of slope at Aravankadu site. The combined effect of surcharge load of building and high pore pressure under heavy rainfall led to intense shearing of slope at Coonoor and Chinnabikatty sites (Chandrasekaran et al. 2013b).

Fig. 9 View of case history sites; **a** slope failure at Aravankadu **b** retaining wall failure at Coonoor **c** slope failure at Chinnabikkaty (Chandrasekaran et al. 2013b)

4.1 Evaluation of Potential of Earthquake Induced Landslides in Nilgiris

As per IS classification of seismic zones, Nilgiris comes under zone III (IS 1893-2002). Though the landsides occurred in this region are rainfall induced, seismicity of this region and past history of earthquakes in and around Nilgiris insists the importance of potential of earthquake induced landslides in Nilgiris. Study on evaluation of potential of earthquake induced landslides in Nilgiris has been carried out by Elayaraja et al. (2015). The peak ground acceleration (PGA) at bedrock level for Nilgiris has been evaluated using deterministic seismic hazard analysis considering a study area of 350 km radius with Ooty as centre. Corresponding to maximum considerable earthquake of 6.8, the PGA at bed rock level is 0.156 g (Elayaraja et al. 2015). Ground response analysis has been carried out on selected locations (seven sites) in Nilgiris by one dimensional equivalent linear method using SHAKE 2000 programme by considering the site amplification due to hilly terrain. From the ground response analysis, it was observed that the PGA of surface motion got amplified to 0.44 g in Ooty site and 0.64 g in Coonoor site compared to 0.39 g of the input motion. The bracketed duration of time history of surface acceleration has increased to 18 s in Ooty site and 20 s in Coonoor site compared to that of 8 s of input motion (Elayaraja et al 2015). Seismic displacement analysis of slope has been carried out for all selected locations using Newmark's method. Ground response analysis and displacement analysis results reveal that out of seven sites considered, Ooty and Coonoor sites have high seismic landslide hazard and other five sites have moderate hazard (Elayaraja et al. 2015).

5 On-Going Detailed Site Specific Geotechnical Investigation on Landslides in Nilgiris

Geotechnical investigation is crucial in landslide studies to understand various causative factors, their failure mechanisms and design suitable site specific remedial measures. A detailed site specific geotechnical investigation is being carried out on landslides in Nilgiris. Different types of landslide (as per Varnes 1978 classification of slope movements) case histories have been identified under this research work. They are debris flow type landslide occurred at Marappalam in 2009 (Fig. 10a), Debris avalanche type landslide occurred at Achanakkal in 2009 (Fig. 10b) and earth slide type landslide occurred at Madithorai location in 2009 (Fig. 10c).

Subsurface investigation has been carried out by making boreholes up to bed-rock and Standard Penetration Test (SPT) has been conducted at every 1.50 m depth. Detailed geophysical investigation has been carried out using Multichannel Analysis of Surface Wave test (MASW) for all three locations. Soil and rock core samples have been collected for laboratory investigation which includes various index and engineering properties of soil/rock samples and determination of residual shear parameters of soil using multiple reversal direct shear test. X-ray diffraction (XRD) analysis and scanning Electron Microscopic (SEM) analysis were performed to identify the mineral compositions of soils and its micro fabric nature. Piezometers have been installed in drilled boreholes to monitor the pore pressure variation under heavy rainfall. An attempt has been made to identify the inter-connection between rainfall (daily and antecedent) and landslide occurrence by establishing correlations between them. Numerical analysis using finite element and limit equilibrium programmes would be performed to understand the failure mechanism and evaluate the factor of safety of the slope. The progressive failure analysis of landslide will be carried out using landslide simulation model LS-RAPID. Different remedial options including geotextiles and reinforcement techniques would be implemented numerically and a suitable option will be suggested which provides better results in terms of safety and stability.

Fig. 10 View of landslides in Nilgiris; **a** Marapalam landslide **b** Achanakkal landslide **c** Madithorai landslide

6 Summary

The present study gives an overview of the various external and internal causes involved in triggering of landslides in Nilgiris. These external and internal causes, either reduce the shearing resistance or increase the shearing stress or affecting both simultaneously. As observed, rainfall is a major triggering factor for landslides in Nilgiris. To dissipate the excess pore pressure generated and reduce the seepage forces during the heavy rainfall, sub surface drainage systems need to be explored. In addition to rainfall, other causal factors such as excavation of slope at toe, loading of slope at crest, blocking of surface drainage system and weep holes of retaining structures, improper planning and design of remedial measures, vegetation removal and waste dumps may also be responsible for triggering of landslides. The awareness of landslides causal factors should be increased among architects, planners, engineers and builders especially in landslide prone areas. It is also observed that the periodic maintenance of surface drainage system and weep holes of retaining structure should be done to keep them clean and reduce the influence of water in triggering of landslides. Excavation of slope at toe is observed in many places especially along transport corridors for widening of roads. Internal slope reinforcement shall be practiced for vertical cut to increase the stability of soil. To stabilize the sliding mass above the failure surface and avoid rotational and translational earth slide, passive piles can also be adopted. In general the remedial measures should be site specific and properly designed and maintained periodically to improve overall stability and safety. The numerical analysis carried out in Nilgiris explains how rainfall and other factors involved in triggering of landslides. Ground response and displacement analysis reveals that Ooty and Coonoor sites have high seismic hazards and other sites of moderate hazard. The on-going detailed site specific geotechnical investigation on landslides in Nilgiris will be useful to understand the failure mechanism of landslides and would help in identifying suitable site specific remedial measures.

Acknowledgements The authors thank Department of Science and Technology (DST-NRDMS Division), Government of India, New Delhi, for sponsoring the study reported in this paper through the project "Geotechnical Investigation on Landslide in Nilgiris district of Tamilnadu" (Project sanction order NRDMS/11/2003/012 Date: 25/09/2014).

References

Anderson, S., and N. Sitar. 1995. Analysis of rainfall induced debris flows. *Journal of Geotechnical Engineering ASCE* 121 (7): 544–552.

Bhandari, R.K. 2006. *The Indian landslide scenario, strategic issues and action points*. India Disaster Management Congress, New Delhi, 29–30 November 2006, Session A2, Keynote address, 1–18.

Cascini, L., S. Cuomo, M. Pastor, and G. Sorbino. 2010. Modelling of rainfall induced shallow landslides of the flow type. *Journal of Geotechnical and Geoenvironmental Engineering, ASCE* 136 (1): 85–98.

Chandrasekaran, S.S. 2010. Assessment of damages induced by recent landslides in Ooty, Tamilnadu. In *Proceedings of the Indian Geotechnical Conference (IGC-2010) GEOtrendz,* IIT Bombay, Mumbai, India, vol. 2, 687–688.

Chandrasekaran, S.S, S. Elayaraja, and S. Renugadevi. 2013a. Damages to transport facilities by rainfall induced landslides during November 2009 in Nilgiris, India. In *Landslide science and practice: Risk assessment, management and mitigation*, vol. 6, 171–176. Rome, Italy: 2nd World Landslide Forum, WLF.

Chandrasekaran, S.S., R. Sayed Owaise, S. Ashwin, R.M. Jain, S. Prasanth, and R.B. Venugopalan. 2013b. Investigation on infrastructural damages by rainfall-induced landslides during November 2009 in Nilgiris India. *Natural Hazards* 65 (3): 1535–1557.

Coppin, N.J., and I.G. Richerds. 1990. *Use of vegetation in civil engineering*, CIRA, Classic House, 174–180 Old Street, London ECIV 9BP, UK.

Elayaraja, S., S.S. Chandrasekaran, and G.P. Ganapathy. 2015. Evaluation of seismic hazard and potential of earthquake-induced landslides of Nilgiris, India. *Natural Hazards* 78: 1997–2015.

Ganapathy, G.P., K. Mahendran, and S.K. Sekar. 2010. Need and urgency of landslide risk planning for Nilgiris District, Tamil Nadu State, India. *International Journal of Geomatics and Geosciences* 1 (1): 30–40.

IMD. 2016. *Nilgiris district rainfall India Meteorological Department.* http://www.imd.gov.in/section/hydro/distrainfall/webrain/tamilnadu/nilgiri. Accessed 01 Aug 2016.

IS 1893–Part I. 2002. *Indian standard criteria for earthquake resistant design of structures part I general provisions and buildings*. New Delhi: Bureau of Indian Standards.

Kourkoulis, R., F. Gelagoti, I. Anastasopoulos, and G. Gazetas. 2011. Slope stabilizing piles and pile-groups: Parametric study and design insights. *Journal of Geotechnical and Geoenvironmental Engineering* 137 (7): 663–667.

Natarajan, T.K., A.V.S.R. Murty, and D. Chandra. 1984. India's first venture relating to subsurface drainage by horizontal drains. In *International Conference on Case Histories in Geotechnical Engineering*, paper 30, Missouri University of Science and Technology, USA.

Nilgiris. 2011. *Natural disaster management—The Nilgiris*. www.nilgiris.tn.gov.in. Accessed 01 Aug 2016.

Popescu, M.E. 2002. Landslide causal factors and landslide remedial options. In *Proceedings of the 3rd International Conference on Landslides, Slope Stability and Safety of Infrastructures,* Singapore, 61–81.

Senthilkumar, V., S.S. Chandrasekaran, and V.B. Maji. 2015. Stability analysis of slopes in Nilgiris district of Tamilnadu, India using strength reduction technique. In *Proceedings of Slope—2015 International Conference*, Indonesia, Bali, 1: I4, 1–7.

Sutejo, Y., and N. Gofar. 2015. Effect of area development on the stability of cut slopes. In *5th International Conference of Euro Asia Civil Engineering Forum (EACEF-5),* vol. 1, 331–337.

Terzaghi, K. 1950. *Mechanisms of landslides*, 83–123. Berkley: Geological Society of America.

Varnes, D.J. 1978. Slope movements and types and processes. *Landslides Analysis and Control, Transportation Research Board Special Report* 176: 11–33.

New Paradigm in Geotechnical Performance Monitoring Using Remote Sensing

Thomas Oommen, El Hachemi Bouali and Rudiger Escobar Wolf

1 Introduction

Measuring lateral ground displacement and vertical settlement have become an integral part of monitoring the performance of geotechnical engineering and design (Mazzanti et al. 2015). Recently, many large infrastructure projects (e.g., transportation corridors, bridges, dams, retaining structures, mines, slope design, and restorations), require the monitoring of displacement and settlement of the structure for its design life. However, the need for improved monitoring has been highlighted by urbanization that has led to the development of areas previously considered geologically unstable and the changing weather patterns. Figure 1 shows the overlay of high susceptibility landslide areas in the continental United States (Godt 1997; Radbruch-Hall et al. 1982) on the densely populated areas and highway network. It is evident from Fig. 1 that several communities and a large portion of the highway network are vulnerable to land-sliding/slope-instabilities.

Traditional geotechnical monitoring techniques require the installation of invasive, in situ instrumentation, such as inclinometers and/or accelerometers to measure rotational displacement and the change in displacement rates, respectively. The installation of this instrument suite, although exceptionally useful (e.g., inclinometers alone can measure landslide movement direction, magnitude, rate, and depth), does have some downsides, including requiring upfront cost of installation and continued upkeep of the instruments, the common accidental improper installation resulting in the under-achievement of actual slope monitoring (Stark and Choi 2008), limited spatial coverage of a single instrument, and alteration of the environment in and on top of the slope.

T. Oommen (✉) · E. H. Bouali · R. E. Wolf
Department of Geological and Mining Engineering and Sciences,
Michigan Technological University, Houghton, MI 49931, USA
e-mail: toommen@mtu.edu

K. Ilamparuthi and R. G. Robinson (eds.), *Geotechnical Design and Practice*,
Developments in Geotechnical Engineering,
https://doi.org/10.1007/978-981-13-0505-4_17

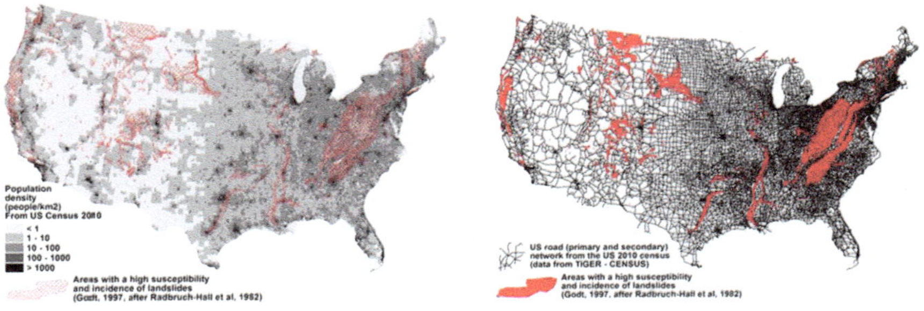

Fig. 1 Left panel shows the population density in the US and the areas with high susceptibility and incidence of landslides. The right panel shows the US road network and the same landslides susceptibility and incidence area as in the left panel. Population and road data from the US 2010 census. Landslide susceptibility and incidence areas from Godt (1997), after Radbruch-Hall et al. (1982)

Particularly, considering the spatial extents of the area that need to be monitored (as shown in Fig. 1), the limited spatial coverage of traditional in situ instrumentation warrants new approach to geotechnical performance monitoring. The objective of this study is to present real world examples demonstrating the application of remote sensing for regional and local scale performance monitoring of infrastructure as a new paradigm that provides high spatial- and temporal resolution for geotechnical performance assessment. In this study we present two examples, (1) a regional scale assessment of settlement in Metropolitan Detroit, MI, and (2) a local scale assessment of a slope displacement at Yukon, Alaska.

2 Regional Assessment Using Synthetic Aperture Radar (SAR)

SAR is an active, side-looking radar system that simulates a large antenna in the azimuth (flight path) direction. This allows for the acquisition of imagery at resolutions higher than that are capable with a stationary antenna. Interferometric SAR (InSAR) is a technique that utilizes multiple radar image acquisitions over the same area to measure temporal changes in satellite-ground distance. Since satellite position is known at each acquisition, this procedure is capable of measuring ground settlement between image acquisitions. Interferograms, generated from a pair of radar images, can accurately measure settlement rates at the cm/year scale. Persistent Scatterer Interferometry (PSI), an InSAR stacking technique that processes more than 20 radar images at once, can accurately measure settlement rates at the mm/year scale (Ferretti et al. 2001; Bouali et al. 2016). The latter is used for this case study.

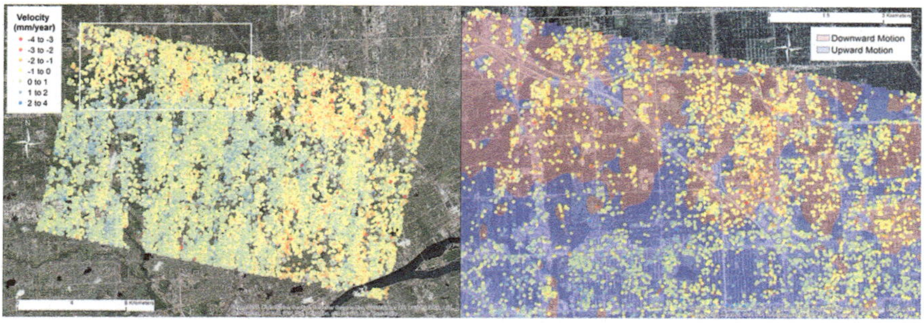

Fig. 2 **a** Left image: ground settlement rates (velocity) from 64,256 PS calculated over metropolitan Detroit, MI between 1992 and 2000 using ERS-1/-2 imagery. Negative values indicate downward motion; positive values indicate upward motion. **b** Right image: close-up view of NW corner (white polygon from Fig. 2a), with nearest-neighbor interpolation of settlement rates showing regions of downward motion (red) and upward motion (blue)

A high concentration of geotechnical assets is located in urban settings, and imagery obtained with remote sensing instruments allows for geotechnical performance monitoring on a regional scale. PSI measures ground settlement rates at thousands of point targets, referred to as persistent scatterers (PS), simultaneously, with data from multiple PS contained within the spatial extent of a single geotechnical asset. The vast amount of ground settlement data calculated using PSI can be seen in the Detroit, MI case study (Fig. 2a). 64,256 PS over 427 km^2 (>150 PS/km^2) were obtained using 50 radar images spanning 1992–2000 from the ERS-1 and ERS-2 satellites. A close-up view of the northwestern extent (white polygon Fig. 2a) is shown in Fig. 2b. PS are draped on an interpolation map showing the spatial extent of downward (red) and upward (blue) areas of motion.

Geotechnical assets in the Detroit metropolitan area in Wayne County, MI were built upon complex geologic conditions due to a variety of glacially deposited sediments and soils. Glacial drift, in the form of glacial lake plains and marginal moraines, rest atop bedrock composed of dolomite, sandstone, and shale. Of the exposed soils in Wayne County, 80.3% are composed of poorly drained, 11.5% exhibit well drained, and the remaining 8.2% are altered, marshland, or open water (Larson 1977). Clay-rich soils, like those found in Wayne County, shrink and swell with variable moisture content and can displace geotechnical assets at the mm-scale. In this example, locations of concern for the structural integrity of geotechnical assets would be the boundaries between downward and upward motion (Fig. 2b), where greatest differential ground settlement occurs. PSI results can then be combined with soil surveys and drainage maps to identify the areas with the greatest risk of hydrologically induced settlement and uplift to infrastructure.

3 Local Scale Assessment Using Photogrammetry

Photogrammetry as a surveying technique has been around for more than a century (Wolf and Dewitt 2000), but only recently have computer power and efficient computational algorithms allowed this method to become a widespread tool for surveying natural and artificial surfaces in the field, at close range (Westoby et al. 2012). Current software processing algorithms allow the use of consumer grade cameras to produce high quality, three dimensional models of the surfaces imaged by the cameras. These models however rely on good camera calibrations and ground control points, to produce precise outputs that represent the surveyed surface. The product can be obtained as point clouds in three dimensional space, similar to three-dimensional point clouds obtained from LiDAR scanners.

In geotechnical engineering, high precision surveying of slopes and other geotechnical assets has been used to monitor the stability of such assets through time. Traditionally this has been done using highly time and other resource consuming methods, involving surveying crews (e.g., total station or precision leveling transects); although the precision achieved with such methods can be very high, they tend to be expensive and only reveal the location of a few points (e.g., benchmarks) on the geotechnical asset of interest. Alternatively, aerial and terrestrial LiDAR scanners can provide a very high density of points with a relatively high location precision (<2 cm), and such representations of the surveyed surface provide a much more detailed and informative model for monitoring purposes, than the sparse points provided by other traditional methods, however the LiDAR equipment and data reduction methods are also expensive and bulky. A less expensive alternative is available through digital photogrammetry, by using consumer grade cameras and cheap aerial (e.g., small unmanned aerial vehicles, or UAV's) or terrestrial platforms, and providing in some cases similar levels of precision.

An example of such applications can be seen at the "Dalton Highway" in Alaska, where the road crosses the Yukon River over a $\sim$750 m long bridge. This bridge serves as a structure for the road and also for the Trans-Alaska Pipeline System. An active landslide has developed some $\sim$100 m NW of the south end of the bridge, raising the concern that the bridge stability could be compromised by the slope stability in the future (Fig. 3).

To monitor if any movement happened on that slope between 2014 and 2015, a helicopter-based aerial survey was undertaken over the landslide area. A 36 megapixel Nikon D800 Digital Single-Lens Reflex (DSLR) camera was used to take between 100 and 150 images of the landslide and surrounding areas. The images were processed with the digital photogrammetry software Photoscan, by Agisoft®. The resulting point clouds had between 8 and 15 million points representing the region of interest. Digital elevation models (DEMs) were interpolated from such clouds, with a ground resolution of 10 cm (Fig. 4, left panel). To assess whether the area had experienced any surface motion between 2014 and 2015, elevation values from the 2014 DEM were subtracted from the elevation values of

Fig. 3 Digital photogrammetric 3D point cloud of the Yukon River Bridge and nearby landslide in 2014

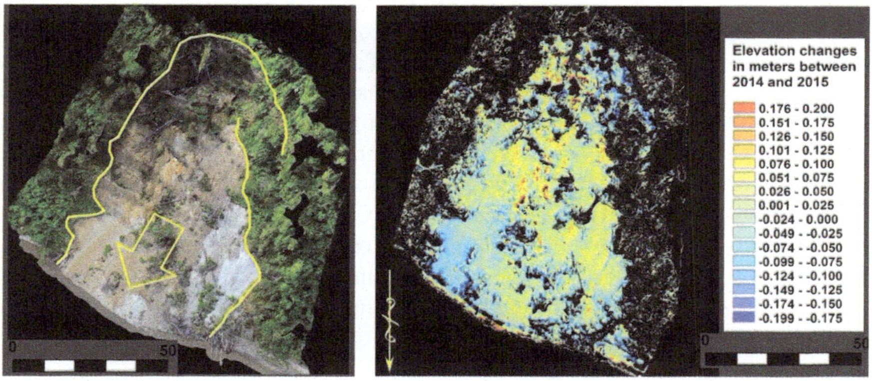

Fig. 4 Left panel shows the landslide in plane-horizontal projection, with landslide scarp and sliding direction indicated. The right panel shows the changes in elevation values for each pixel in the DEMs, from 2014 to 2015

the 2015 DEM (Fig. 4, right panel). The result shows that elevation changes were subtle, with many elevation changes happening over only a small area, and being associated with the movement of individual boulders or pieces of debris (e.g., tree trunks). A more regional trend of downward motion was observed at the lower end of the slide, and could be related to bank erosion at the foot of the slide, the values of such vertical change were however modest (<10 cm). Overall it seems that very little displacement related to landslide activity happened at the site between 2014 and 2015, a somewhat relieving news from a hazard perspective, for those managing the Dalton Highway and the Trans-Alaska Pipeline.

4 Discussion and Conclusion

Remote sensing can be used to monitor geotechnical assets through time, providing key information about the surface displacement and deformation, which can then be used to assess the asset's stability, or its reduction through time. Satellite based InSAR allows the relatively quick monitoring of large areas with very high precision, with a millimeter scale precision for individual displacements, or velocity precision on the order of a few millimeters per year. This is particularly the case for areas with abundant infrastructure, e.g., urban areas, and transportation corridors, as human built objects tend to reflect the radar signal more consistently through time, allowing for stacking methods, like PSI to give better results in such areas. For some applications, the data-point density given by satellite based InSAR methods like PSI may not be high enough, but the information on surface displacement can be used to focus the efforts for higher point density collection through other methods.

Local scale assessments and monitoring of surface deformation for geotechnical assets can be done using digital photogrammetric methods, based on relatively cheap DSLR cameras, and user-friendly processing software. The precision achieved through such method is on the order of a few cm, and can be used to detect large scale displacements of landslides and similar unstable geotechnical assets. The precision and degree of detailed is comparable to more expensive LiDAR scanning methods that are usually applied to this kind of geotechnical problems. The availability of cheap, lightweight and compact sensors (i.e., photographic cameras), and platforms (e.g. small UAVs for aerial deployment, and vehicle or pedestrian use for terrestrial deployment) makes this method very versatile. The need for ground control points is however important to keep in mind, as the quality of the end product, e.g. the surface displacement over time, may depend on such ground control.

Acknowledgements This project was funded by the U.S. Department of Transportation (USDOT) through the Office of the Assistant Secretary for Research and Technology (Cooperative Agreement No. RITARS-14-H-MTU). Alyeska Pipeline Service Company for access to helicopter for data collection. ERS-1 and ERS-2 SAR data provided by the European Space Agency.

References

Bouali, E.H., T. Oommen, and R. Escobar-Wolf. 2016. Interferometric stacking toward geohazard identification and geotechnical asset monitoring. *Journal of Infrastructure Systems* 22 (2): 05016001. https://doi.org/10.1061/(asce)is.1943-555x.0000281.

Ferretti, A., C. Prati, and F. Rocca. 2001. Permanent scatterers in SAR interferometry. *IEEE Transactions on Geoscience and Remote Sensing* 39 (1): 8–20.

Godt, Jonathan W. 1997. Digital compilation of landslide overview map of the conterminous USA. *Report 97-289*: http://landslides.usgs.gov/hazards/nationalmap/.

Larson, J.D. 1977. *Soil survey of Wayne county area, Michigan*. United States Department of Agriculture Soil Conservation Service in cooperation with Michigan Agricultural Experiment Station, 94 p.

Mazzanti, P., F. Bozzano, I. Cipriani, and A. Prestininzi. 2015. New insights into the temporal prediction of landslides by a terrestrial SAR interferometry monitoring case study. *Landslides* 12 (1): 55–68.

Radbruch-Hall, D.H., R.B. Colton, W.E. Davies, I. Lucchitta, B.A. Skipp, D.J. Varnes. 1982. Landslide overview map of the conterminous United States. US Geological Survey. *Professional Paper 1183. Digital Version of the Map Created by Godt JW and Found as an Open-File Report 97-289*. http://landslides.usgs.gov/hazards/nationalmap/.

Stark, T.D., and H. Choi. 2008. Slope inclinometers for landslides. *Landslides* 5: 339–350.

Westoby, M.J., et al. 2012. 'Structure-from-motion' photogrammetry: A low-cost, effective tool for geoscience applications. *Geomorphology* 179: 300–314.

Wolf, Paul R., and Bon A. Dewitt. 2000. *Elements of photogrammetry: with applications in GIS*, 696. New York: McGraw-Hill.

Characterisation of Vesicular Basalts of Mumbai Using Piezoceramic Bender Elements

A. Juneja and M. Endait

1 Introduction

Elastic wave velocity is not only useful for geophysical and seismological studies, but also for rock characterisation. The wave velocity is related to rock's stiffness, strength and many other intensive properties such as density and porosity. Practical engineering problems can be solved (e.g. drilling and blasting, stability assessment of tunnels and engineering seismology) if the characteristics of reflected (or refracted) waves are known. One of the important elastic constant determined from the elastic wave, is the small strain shear modulus, $G_{\max}$. $G_{\max}$ is occasionally referred to as the "dynamic" shear modulus in rock engineering, because it is determined using non-destructive cyclic testing methods. Shear wave velocity, v_s and the density, ρ of the geomaterial, are utilised to obtain $G_{\max}$ from the equation

$$G_{\max} = \rho v_s^2 \tag{1}$$

It is easy to generate and measure the compression (or the primary) wave velocity, v_p in the field, which however, is not true for v_s (Wang et al. 2009). Therefore, many a times, it can be tempting to use elastic relationships between v_s, v_p, and Poisson's ratio, n to replace v_s in the above equation. Unfortunately, these relations can sometimes lead to unrealistic value of n, especially when the rock contains significant amount of impurities (Diamantis et al. 2009). Because both

A. Juneja
Department of Civil Engineering, Indian Institute of Technology Bombay, Powai, Mumbai 400 076, India
e-mail: ajuneja@iitb.ac.in

M. Endait (✉)
Department of Civil Engineering, Sandip Institute of Engineering and Management, Nashik 422213, India
e-mail: mahesh.endait@siem.org.in

K. Ilamparuthi and R. G. Robinson (eds.), *Geotechnical Design and Practice*, Developments in Geotechnical Engineering,
https://doi.org/10.1007/978-981-13-0505-4_18

Table 1 .

Reference	Relation	Rock type
Carroll (1969)	$v_s = 0.29{v_p}^{0.82}$	Volcanic
Christensen (1977)	$v_s = 0.59{v_p}^{-0.32}$	Basalt
Brocher (2005)	$v_s = 0.7858 - 1.2344v_p + 0.7949{v_p}^2 - 0.1238{v_p}^3 + 0.0064{v_p}^4$	Earth crust
Diamantis et al. (2009)	$v_s = 0.52v_p$	Serpentinites
Wang et al. (2009)	$v_s = 0.622{v_p}^{-0.382}$	Basalt
Wadhwa et al. (2010)	$v_s = 1.1{v_p}^{0.92}$	All type of rocks

v_p and v_s are affected by the structure, composition and intensive properties of the rock, they may not necessary have unique relationship, which otherwise, can be derived from the theory of elasticity. It is therefore prudent to relate v_p and v_s to the rock quality (Brocher 2005; Wang et al. 2009), because of which empirical relations between v_s and v_p are used. Nevertheless, not all properties of the basalt have been established, except for a few studies (Christensen 1977), which confirm that the elastic wave velocity varies within the different basaltic formations. Table 1 shows a few common relations between v_s and v_p for different rocks. It is interesting to note that flexible and high order polynomial fits have been used for many rocks to achieve a reasonably acceptable coefficient of variance (CoV).

In the laboratory, ultrasonic pulse test is one of the most commonly used method to help determine v_p in rock cores. The use of piezoceramic bender element to estimate v_s (and corresponding G_{max}) in the laboratory is also truly convenient.

These elements are made up of piezoelectric-ceramic material sandwiching a thin metal plate. Shirley and Hampton (1978) were perhaps the first to report the use of bender elements in soil laboratory testing. Dyvik and Madshus (1985) estimated that the strain of the transmitting bender element was in the range of 10^{-5}. Leong et al. (2005) estimated that the strain of the receiving bender element was in the range of 10^{-6}. Because of these small yet finite deformations of the elements, most of the laboratory works has been limited only to soils.

These also appear to be a significant gap in the limited data obtained with bender element testing in hard rocks (Arroyo et al. 2010). Adaption of the piezoceramic bender elements in comparatively stiffer material (such as rock) will make this test more versatile.

A Study of elastic wave propagation across the joint filled with and without gouge material has importance in practical field conditions. Joint thickness, properties of gouge material, joint roughness coefficient and normal stress on joints are some of factors on which wave propagation characteristics are depend. Amongst these characteristics joint orientation is an important factor which is less explored.

In the field, estimation of v_s is difficult, and the same is then estimated through empirical relationship with v_p. The objectives of the present study are, to supplement the relationship between v_p and v_s with the new data for vesicular basalt rock and to study the effect of joint orientations on v_p and v_s.

Rock cores were collected from over 22 locations across the Mumbai city. This region is known to be a part of the western Deccan volcanic province and is the youngest rock of the Eocene age. The grey Basalt rocks vary from friable, highly weathered to very strong and compact. Much of the amygdaloidal rocks are filled with zeolites, calcites and other secondary silica bearing minerals. At places the Basalt rock shows porphyritic texture. NX-size cores were obtained from boreholes drilled using the double tube rotary drilling method. The cores were tested in the laboratory using piezoceramic bender elements, which were fabricated to estimate v_s in basalt rock.

Statistical analysis of the laboratory test results was conducted using a curve fitting tool (Matlab 2011). The correlations obtained in this work were then compared with other relations, which are available in the literature.

2 Measurement of v_p

v_p was measured using an ultrasonic pulse velocity tester (Pundit model PL-200). The emitter and the receiver were set at 54 kHz. The equipment was calibrated to read v_p of standard aluminium cylinder.

In order to isolate the size-effect from the shape of the waveform, the size of the chosen cylinder was the same as that of the rock cores. In this procedure, the transducers were clamped onto the two polished ends of the cylinder. A high voltage pulse was then generated by the emitter and transmitted through the cylinder. v_p was calculated using the time taken by the pulse to travel from the emitter to the receiver.

The instrument was tuned to read v_p of about 6321 m/s, which resulted in the n of aluminium equal to that obtained from the theory of elasticity (Arroyo et al. 2010). v_p of the dry rock cores was determined in the similar manner.

3 Measurement of v_s

v_s was determined using bender elements. The piezoceramic elements were fabricated using lead zicronate titanate based material, SP-5A. It corresponds to US Department of Defence, Navy type material-II. A small 0.15 mm thick brass sheet

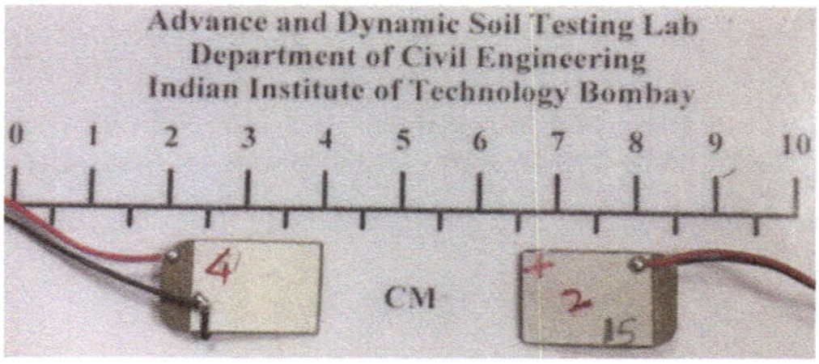

Fig. 1 Piezoceramic bender elements

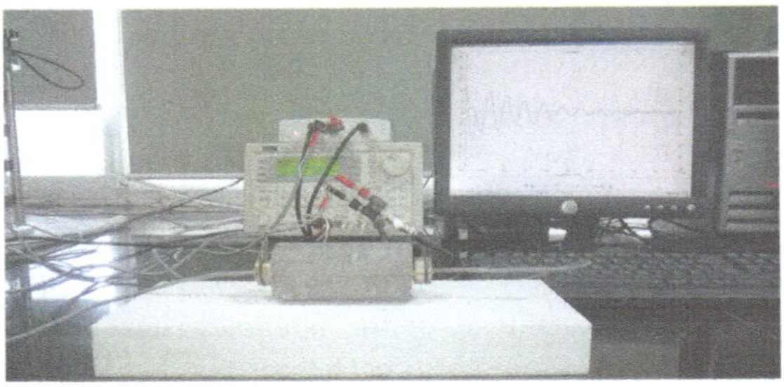

Fig. 2 Setup to measure v_s in basalt rock

was rigidly bonded on its two sides by 16 mm × 12 mm × 0.25 mm piezoceramic plates. The bender elements used in this study are shown in Fig. 1.

Figure 2 shows the setup to measure v_s in basalt rock. Thin slots were drilled on either sides of the core and filled with filler material. Transducer plates were then inserted into the slots and their housing clamped on the two cylindrical surfaces. With this arrangement, direct contact between the rock core and the transducer plates was prevented to avoid damage to the transducer. The sample was then placed over a 45 mm thick geofoam to help absorbed and prevent rogue waves from reflecting back into the sample.

The transmitter transducer was excited by the waveform generator (Aim-TTi Model TGA 1241) to produce 80 kHz sinusoidal pulse of 20 V amplitude. The sampling rate of the generator was set at 40 MHz sampling frequency and 12 bit vertical resolution. The use of a single wave simplified the analysis by eluding the phase velocity and the group velocity from the shear wave velocity (Leong et al. 2005).

Wavelength was less than half the sample length, to prevent near-field effects. An oscilloscope (Picoscope model 4824) was connected to the receiver transducer to pick electric signals. The oscilloscope was connected to both the function generator and the receiver to simultaneously record the transmitted and received signals.

The sampling rate of the oscilloscope was set at 80 MHz and 12 bit resolution. These signals were sent to a computer for processing using a high-speed data acquisition card.

4 v_p and v_s in Jointed Rock

A few rock samples were sliced using a diamond cutting disc to produce joints at 0–50° to the horizontal. The blocks were clamped to measure v_p and v_s. The dry joints created by slicing the samples had zero joint roughness coefficient obtained by comparing the roughness profile provided by Barton and Choubey (1977).

Fig. 3 Jointed rock samples with weak gouge material (kaolin clay)

To simulate weak joint condition, the block separation was filled with 1 mm thick kaolin slurry at its plastic limit. Figure 3 shows the rock samples with gouge material prepared for this study.

5 Results and Discussion

More than 115 ultrasonic and bender element tests were conducted. v_p ranged between 1.89 to 5.89 km/s and v_s between 0.94 to 3.31 km/s. γ_{dry} of the tested samples varied from 19.03 to 29.14 kN/m^3 and n was observed to vary from 0.11 to 29.8%.

The elastic moduli, E calculated using the theory of elasticity ranged from 5.5 to 72.3 GPa. G_{max} and bulk modulus, K were in the range of 2.1–24 and 5.6–57.9 GPa, respectively.

Figure 4a, b show selected bender element test results. First direct arrival method in time domain was used to determine arrival time. The vertical arrow in the

Fig. 4 (**a**, **b**) Transmitted and received signals in basalt samples in Test No.: **a** 1; **b** 2

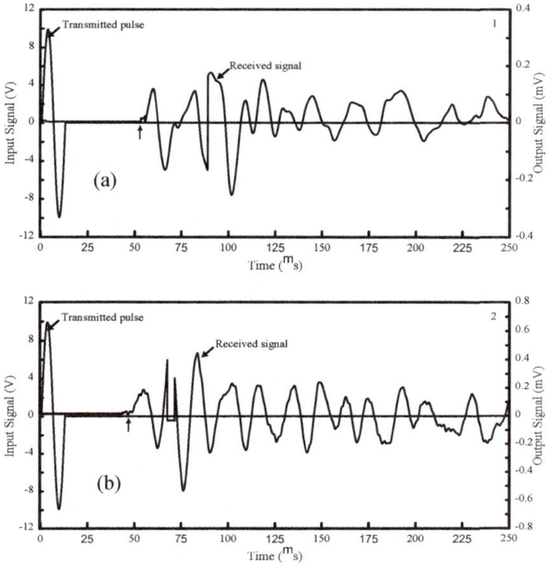

figures indicates the time at which the first positive signal was recorded. In this time, the wave travelled from the tip of the transmitter to the tip of the receiver and, was taken equal to the travel time (Leong et al. 2005).

The tip-to-tip distance between the transmitter and the receiver was 88 mm. As can be seen from results, near-field effects were absent in the received signal, which made it easy to captured the first positive deflection in the curve.

Similar observations were made by Arroyo et al. (2010). No cross talk effect was observed in the received signal since the bender elements were properly grounding to the waveform generator.

5.1 Relation Between v_s and v_p

Figure 5 shows the variation of v_s with v_p. The figure also shows the relations obtained from the literature. As can be seen, there is a reasonable agreement between the present data and the published correlations. However there are also some differences. Wadhwa et al. (2010) tends to over-predict v_s.

On the other hand, Carroll's (1969) findings largely tend to under-predict v_s. Wadhwa et al. (2010) and Carroll (1969) obtained v_s from field test results, albeit Wadhwa et al. (2010) used the shallow rock data.

Basalt in this study was mostly unweathered, with its amygdale cavities filled with secondary minerals such as zeolites, calcites and silicates.

Therefore it is unlikely that this basalt was porous which would otherwise have affected the observed v_s by the high overburden pressures used in Carroll's (1969) results.

The strongly trended variation of the results in Fig. 6 suggests that the discrepancy is deterministic rather than stochastic and is unlikely to be due to the experimental errors.

The variation of v_s and v_p can therefore be determined directly by fitting an intuitively acceptable curve to the data as shown in the figure. The equation of this curve is written as

$$v_s = 0.53 v_p \quad (2)$$

Equation (2) yields R^2 of 0.81 with SSE equal to 4.927 and RMSE equal to 0.2.

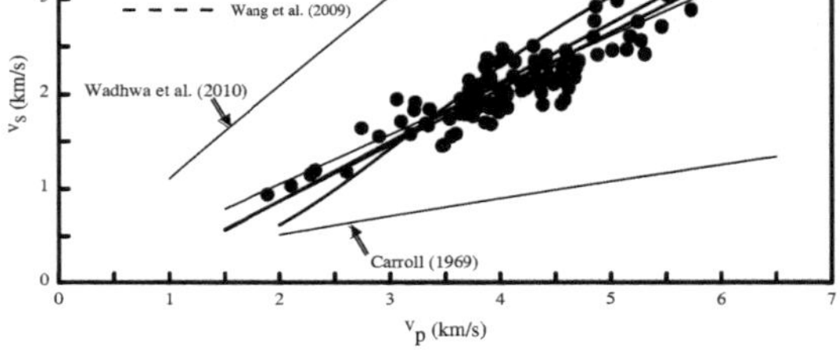

Fig. 5 Relationship between v_s and v_p

5.2 *Effect of Jointed Rock on* v_s *and* v_p

Jointed rock samples with different joint orientation were examined for v_p and v_s. In this case, the samples was cut into two pieces to create one joint. Kaolin slurry was used to lubricate the joint, which otherwise was held dry in half of the tests.

Figure 6a, b show transmitted and received signals in jointed rock. Clearly, the velocity and amplitude of the propagating wave changed due to existence of the joint. The figures also include the output signals when the rock was intact. The vertical arrow in each set, shows the time at which the first positive signal was recorded.

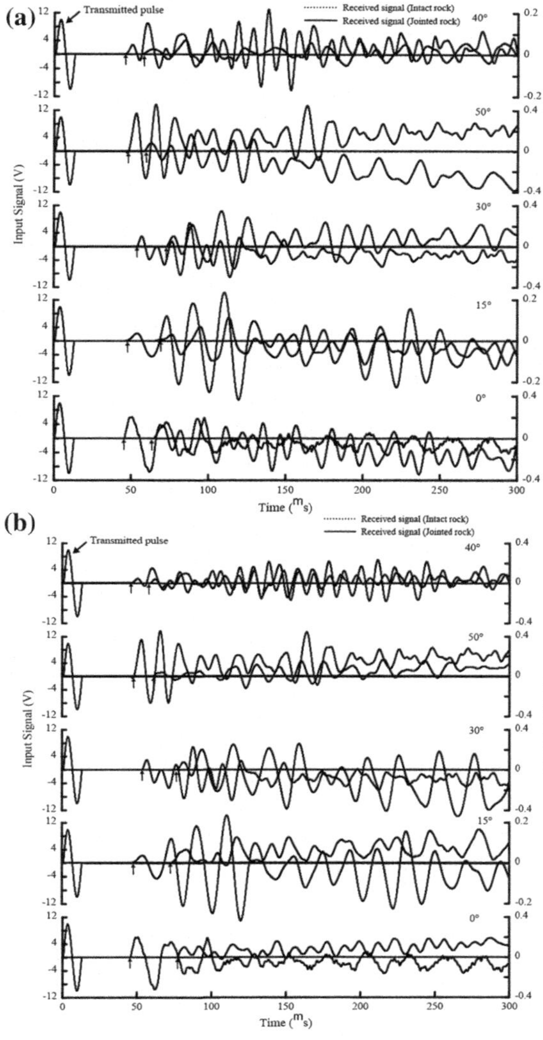

Fig. 6 Transmitted and received signals in jointed rock: **a** filled joint; **b** dry joint

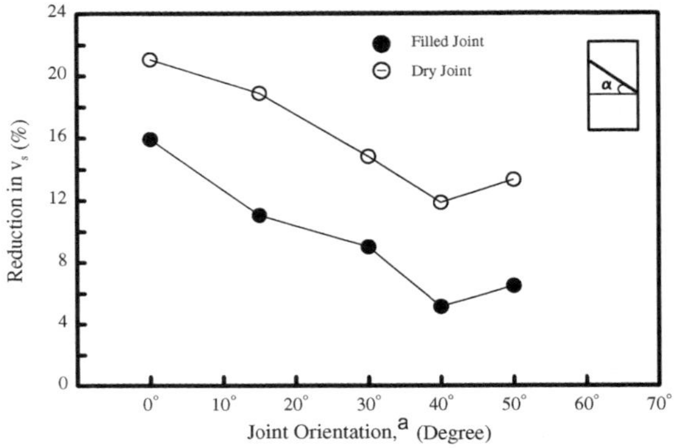

Fig. 7 Effect of filled and frictional joint orientations on v_s

It can be observed that, the presence of joint definitely increased the travel time. In case of filled (lubricated) joint, the travel time increased by as much as 16% while for dry joint, the increase was 21% when the joint was horizontal (90° to the vertical).

Interestingly, the amplitude of the received signal also decrease due to the energy loss in the joints.

Figure 7 shows v_s of the jointed rock. Similar to the above observations, v_s was also affected by the joint orientation. v_s was at its maximum when the orientation was 40° and reduces continuously and attains its minimum when the joint was horizontal.

In filled joint, the reduction in v_s was one-half of the dry joint. This was surprising because the dry joint should have offered greater resistance to shear due to the interlocking affect.

The reason for this discrepancy is not clear. One possible reason for this abnormality is that, the dry joint was made by slicing the sample with a cutter which produced surfaces of joint roughness coefficient (JRC) equal to 0-to-2.

Possibly this created a smooth surface, which also attracted air-pockets within the joint, although the two rock pieces were tightly clamped together.

Figure 8 shows the reduction in v_p in jointed rock. It can be seen from the figure the percentage reduction in v_p for both filled and frictional joints are tighten within say 1 and 4%. In this case, the direction of P-wave was normal to the joint. In all the cases, the same v_p was recorded.

6 Conclusion

Properties of vesicular Basalt were investigated in this study. Basalt rock cores were collected from 22 different locations across Mumbai city.

Bender element test was used to determine the v_s in rock with some modification in core samples.

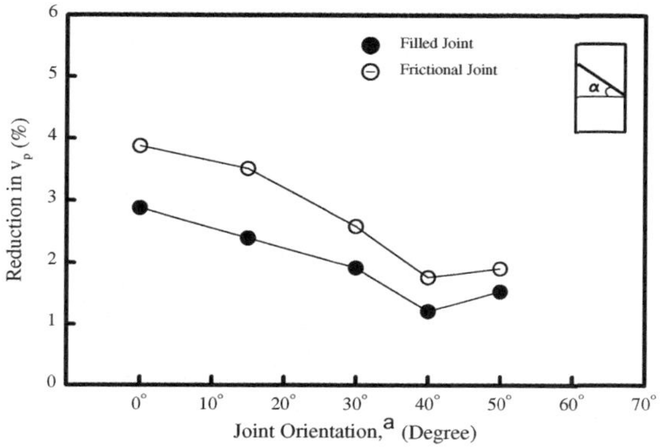

Fig. 8 Effect of filled and frictional joint orientations on v_p

The tests were conducted with the use of appropriate filler material to seal the gap between the piezoceramic plates and the rock core. The first arrival method was used to estimate the arrival time of the elastic wave because the near-field effects were absent and a sharp first positive deflection could readily be captured in all the tests.

v_p and v_s were linearly correlated with a high degree of precession.

The effects of joint orientation filled with and without gouge material on the velocity of seismic waves was studied. Results showed that v_s and v_p are affected because of the presence of joint, as expected.

The wave velocity decreased as the joint angle decreased 50°–0° that is, the maximum reduction in the wave velocity occurred when the joint was horizontal. Primary waves are less influenced by the orientation of the joint.

The presence of gouge material in joints also alters the wave velocities.

It was observed that v_p and v_s further reduced in dry joints when compared to lubricated joints with filler or gouge material.

This interpretation, is in fact, unwarranted because the dry joints were smooth and planar, which otherwise can yield higher results for more rough joint surfaces.

References

Arroyo, M., J.A. Pineda, and E. Romero. 2010. Shear wave measurement using bender elements in argillaceous rocks. *Geotechnical Testing Journal, ASTM* 33 (6): 1–11.

Barton, N.R., and V. Choubey. 1977. The shear strength of rock joints in theory and practice. *Rock Mechanics* 10 (1): 1–54.

Brocher, T.M. 2005. Empirical relations between elastic wave speeds and density in earth's crust. *Bulletin of the Seismological Society of America* 95 (6): 2081–2092.

Carroll, R.D. 1969. The determination of the acoustic parameters of volcanic rocks from compressional velocity measurement. *International Journal of Rock Mechanics and Mining Science* 6 (6): 557–579.

Christensen, N.I. 1977. Seismic velocities and elastic moduli of igneous and metamorphic rocks from the Indian Ocean. In *Indian Ocean geology and biostratigraphy*, ed. J.R. Heirtzler, H.M. Bolli, et al., 279–299. American Geophysics Union.

Diamantis, K., E. Gartzos, and G. Migiros. 2009. Study on uniaxial compressive strength, point load strength index, dynamic and physical properties of serpentinites from Central Greece: Test results and empirical relations. *Engineering Geology* 108 (3–4): 199–207.

Dyvik, R., C. Madhus. 1985. Lab measurement of Gmax using bender elements. In *Proceedings ASCE Annual Convention: Advances in the Art of Testing Soils under Cyclic Conditions*, 1–7.

Leong, E.C., S.H. Yeo, and H. Rahardjo. 2005. Measuring shear wave velocity using bender elements. *Geotechnical Testing Journal, ASTM* 28 (5): 1–11.

Shirley, D.J., and L.D. Hampton. 1978. Shear wave measurements in laboratory sediments. *Journal of Acoustic Society of America* 63 (2): 607–613.

Wadhwa, R.S., N. Gosh, and S. Rao. 2010. Empirical relation for estimating shear wave velocity from compressional wave velocity of rocks. *Journal of Indian Geophysical Union* 14 (1): 21–30.

Wang, Q., S. Ji, S. Sun, and D. Marcotte. 2009. Correlations between compressional and shear wave velocities and corresponding Poison's ratios for some common rock and sulphide ores. *Tectonophysics* 469 (1–4): 61–72.

Geotechnical Challenges and Opportunities in Large Infrastructure Projects—Learning from Failures

Jaykumar Shukla and Soumen Sengupta

1 Introduction

The correct understanding of the anticipated loading conditions during entire life cycle of structure (during construction and in-service conditions) is an important aspect of designing large industrial structures. The structures designed for loading considering in-service conditions may experience distress during construction conditions and small mistake during the construction can stall project execution and incur heavy cost and schedule impact. Present paper shares recent experience highlighting some of the failures observed especially during construction phase, thereby emphasizing need to understand the whole life cycle from construction to in-service for designing the structures in better way. It also highlights the need for understanding the soil–structure interaction failing to which project may experience cost and schedule overrun.

2 Deep Excavations and Underground Structures

The scale pit (46.05 m × 19.6 m) was planned with bottom of base raft located at −24 m from ground level. The site area is fairly leveled having average elevation of +5 m above mean sea level. The strata comprises essentially of silty fine to medium sand with an average permeability of 10^{-4} m/s up to explored depth of −25 m. No bed rock or impervious layers has been encountered. The ground water

J. Shukla (✉)
L&T Sargent & Lundy Limited, Vadodara, India
e-mail: Jaykumar.Shukla@Lntsnl.com

S. Sengupta
Civil Structural and Architectural Department, L&T Sargen & Lundy Limited, Vadodara, India

K. Ilamparuthi and R. G. Robinson (eds.), *Geotechnical Design and Practice*, Developments in Geotechnical Engineering,
https://doi.org/10.1007/978-981-13-0505-4_19

table is about 1–1.5 m below ground level. The construction of pit was done initially by sinking and grabbing method just like construction of well foundation. After well lowered up to specified depth, execution team has planned to carry out underwater concreting to plug its bottom and pumped 500 m^3 concrete in the well using tremie without dewatering. After putting underwater concrete, verification by divers and sufficient waiting period of strength gain, the site team started dewatering from inside the well using direct pumping assuming that bottom is plugged. Direct removal of water within the scale pit created gradient between the outside ground water level (at 1.5–2 m below natural ground level) and water level within the scale pit. This resulted in large uplift water pressure at the base raft which subsequently blown up entire unreinforced base raft. Along with seepage within the pit, large quantity of silt and fine sand have entered inside the scale pit from the bottom of the pit thereby causing subsidence close to the pit (see Fig. 1). Due to excessive seepage into the pit from bottom along with silt and sand has resulted settlement of structures in the vicinity and the commissioning of the other structure was temporarily stalled. The construction of the scale pit was delayed as without plugging the bottom other operations cannot be started. All these consequences severely affect the project progress of surrounding structures.

It was decided to dewater in very controlled manner using 9 nos of 20BHP pumps installed in the outer periphery of pit from the depth of 30–35 m. The peak discharge of all the pump together is about 150 m^3/h. Settlement of the structures were monitored by installing 10 points on the entire piled raft by survey instruments to counter any alarming situations.

Dewatering through all the wells were continued till the completion of the construction of the bottom raft along with some other beams to take the uplift pressure (see Fig. 2). It took more than 6 month to restore the construction of the scale pit and commissioning of the surrounding structure experience Later backfilling of the wells were carried out using gravels and sand. As the water table is

Fig. 1 Collapsed portion of ground around scale pit

Fig. 2 Scale pit after restoring the base raft

Fig. 3 Collapsed and slide of the side earthen embankment toward the deep excavation

located at −2 m from ground level, the grouting was also suggested for the concrete joints. For more detailed information on the issue and restoration, readers are requested to refer Shukla et al. (2009).

In another project instance, a pump house requires to be installed very close to perennial river in Bangladesh. The river is only 50 m away from the facility area. The subsoil conditions are silty fine sand to clayey silt. Excavation need to be carried out up to (−) 22 m from the natural ground level and contractor decided to go for open excavation to place the bottom raft of the pump house. The area partially excavated using open excavation and when the seepage water could not be controlled, contractor went ahead with sheet pile enclosure surrounded by bamboo wall. However, the water level in the river was rising as monsoon is approaching very fast. Site started dewatering from the pit in exigency of completing the excavation.

Few days later, in the night, one side of the system gave up and under the action of toe collapse, entire excavation on one side was collapsed (Fig. 3) and water rushed in the pit all of the sudden. It was fortunate that incident occurred in the night and no human workforce present in the pit which somehow prevented causality. Later, the entire installation has to stop for almost 8 months and when water level recedes in the river, the restoration took place. It was clear indication that in exigency, the dewatering started from the inside actually triggered the seepage and toe failure.

3 Buried Pipe Structures

Considering the amount of water to be circulated for large power plant units, the diameter of cooling water pipe was selected 3840 mm (149.5 in.) for one of the supercritical power project installations. These pipes, designed to operate at a normal pressure of 5 bar, were buried with 1.5 m earth cover without any encasement to facilitate faster site construction. Also, the construction time span of such pipes were as high as 2–3 years and during this period, the empty pipes were exposed to different seasons and construction loadings. Since, these pipes were laid at almost same time, the operating conditions varied based on need for different units i.e. may not be operating at the same time. The excavation for placing four (4) numbers of CW Pipes was commenced from May 2011 onwards. Placing of cooling water pipe line was commenced from July 2011 and completed by January 15, 2012. It was reported that supply and return pipes were remain empty up to July 14, 2013. Hydro test of supply line commenced from July 14, 2013. On early hours

Fig. 4 Collapsed area of buried pipe apx 150 m of pipe along wall was collapsed

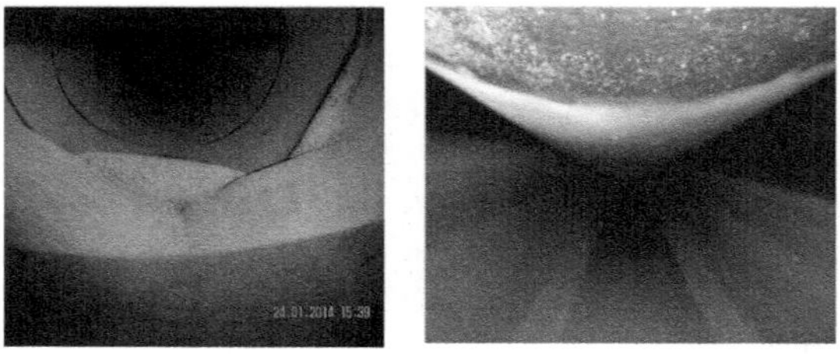

Fig. 5 Photographs of inside of the pipe; snap through failure is visible

of July 24, 2013, a severe soil subsidence at return line along with heave was reported to take place catastrophically. Figures 4 and 5 illustrates the observed failure for the pipeline stretch. It was informed that a 4.5 m wide parallel trench was excavated along the pipeline stretch in order to construct a boundary wall between the pipeline and switchyard. Excavation of trench was commenced in October 2012 (by some other contractor). The pipe along with the soil above collapsed downwards locally and was continued for about 150 m on either side. The excavation for boundary wall was carried out up to a depth up to center-line of pipe. The excavation was filled with storm water as well. On further inspection of pipeline from inside, it was observed that the maximum down ward deflection of the pipe was about 2.0 m at failed segment. It was informed that the downward deflection is local across the cross section of the pipe and mostly restricted to top portion of the pipe only thereby indicating “snap through failure” of pipes. The balance portion of the cross section has retained the circularity more or less. It was also informed that the bottom of pipe has not undergone any significant deflection.

Shukla et al. (2016) carried out detailed analysis to investigate the failure causes and it was recommended that due to parallel trench excavation along the pipe installations, created differential soil support conditions which are required for supporting mechanism of large diameter buried flexible pipes. The entire 150 m stretch of the pipeline was replaced by new pipe installations. Significant cost overrun was experienced and project was delayed since CW pipe is key element for process cycle and without it power plant operation cannot commence. Readers are requested to refer Shukla et al. (2016) for more detailed analysis and greater insight of actual problem.

In another instance, the empty pipes were laid and due to some construction requirements, the top earth cover (soil above pipe) was excavated temporarily. However, flooding due to sudden storm flooded the excavated portion and laid pipe experienced the hydrostatic uplift due to this temporary flooding. However, due to insufficient earth cover over pipe, pipes got uplifted in some section, see Fig. 6.

Fig. 6 Buried pipe experienced distress due to floatation

This instance indicates the power of hydrostatic uplift as it could bend and uplift 3.8 m dia steel pipe (with 24 mm wall thickness). The incident caused damaged to already laid pipes in a segment which later require to be replaced. The project though do not experience schedule delay, but suffer heavy cost for rectification and replacement of the pipe segment.

4 Containment Structures (Reservoir and Dykes)

In most of the industrial developments, containment structures are very routine to store water, i.e., raw water reservoir or waste materials, i.e., ash ponds. For such facilities, generally HDPE or LDPE linings are used to provide impervious barrier. These geosynthetic linings are further specified to be protected using precast tiles or brick laid over geosynthetic. While slope stability studies are performed for almost all such facilities, "Veneer Stability" of such protection lining laid over smooth geosynthetic are seldom performed. In order to save the parcel of land used for the embankment formation, the steep side slopes are decided based on the slope stability studies and tiling or protection lining resting on steep slopes usually fail under "Veneer" sliding, see Fig. 7. Additionally, on high slopes, the absence of anchorage of laid geosynthetic at stages, creates additional problem. Any movement of laid geosynthetic opens up the joints in the tiles which subsequently triggers veneer sliding.

In another instance, the designer fail to recognize the permeability of founding stratum and dyke was resting on comparatively pervious stratum. It leads to seepage which emerged out beyond the toe of dyke eroding ash particles from the dyke body, see Fig. 8. The remedial measures are even costlier and plant operation was under pressure as the containment premises cannot be used unless the restoration takes place.

Fig. 7 Sliding of thin protection tiles over smooth geosynthetic laid over steep slopes

Fig. 8 Seepage of ash beyond the toe of embankment as dyke is resting on pervious stratum

Fig. 9 Lateral sliding of soil due to rapid filling of iron ore slock

5 Other Issues

It is important for a designer to recognize the interaction effect envisaged by adjacent structure and failing to which can create troublesome situations. In one of such situation is illustrated in Fig. 9. The stacker reclaimer foundation was designed considering the pile foundation at certain spacing. However, designer failed to realize that adjacent area will be used for stacking iron ore and rapid filling of iron ore can exert lateral earth pressure against the reclaimer foundation. The ore stack height is 19 m and during the first filling, the entire area stack yard was filled in one go. As shown in the Figs. 9 and 10, the top soil (clay soil) was slided outwards and exert the reclaimer foundation and other utilities running parallel.

Fig. 10 Closure look of lateral sliding against the reclaimer foundations

Though the reclaimer foundation was built over mass concrete resting on pile foundations, the pressure was in significant intensity which could lift the utilities, i.e., Fire pipes, drains, etc., which got uplifted. This was arrested by immediately removing the iron ore stack load and then ground was improved using discrete stone columns in order to present any sliding induced pressure on the structures adjacent to the stock yard. It is also observed that the exerted pressure was such huge that it has misaligned the reclaimer foundation at few places. It also demonstrated that piles on which reclaimer foundations are resting can be subjected to significant lateral loading (across the crane movement) and designer must consider such scenarios while designing the foundation close to such stock yards.

It is very helpful to build the 3D model in the design office and check the relevant interference with the adjacent structure. This is more relevant to the underground utilities and structures and more important when the agencies constructing the structures are different. In one of such plant constructions, designer failed to understand the actual construction sequence and planned a deep sitting pipeline adjacent to the planned pipe rack stretch. The interference was checked in the model for actual relative positions of the two structures, i.e., pipeline and pipe rack foundations and gave clearance that structures are inference free and can go ahead for construction. Since the deeper structures shall be constructed first, then an agency was mobilized to lay pipe lines.

For buried structures, backfilling is seldom of any significance. The pipe was laid in a very narrow trench and after laying of pipe, the area was backfilled with the excavated soil. However, the compaction was not proper and it was loose dumping of soil over the laid pipeline see Fig. 11. Later the pipe rack foundations were laid assuming that the foundations are resting on the virgin soil and concrete was placed for many foundation units. The backfilled mass got settled subsequent to rain induced flooding in the area. The foundations experience severe tilting and settlements for almost all the pipe rack foundations see Fig. 12. It was fortunate that only foundation raft was cast and raising he columns/pedestals were not in place. Later

Fig. 11 The pipeline laying and backfilling relative to the pipe rack foundation

Fig. 12 Tilted foundations of pipe rack

these foundations were recommended to be abandoned and new pipe rack route was decided away from the excavation footprint of the trench.

6 Recommendations

There may be many factors contributing to a failure. However, most common is ignorance and lack of overall vision of the system a whole. For example, while excavating and dewatering, in the exigency to finish the work or to see the bottom of raft, the interaction effects are somewhere missed which leads to a catastrophic event. Once experienced the failure or distress, the remedial measures are even more cost inviting and stalling the project which was experienced in few of the cases, had even subsequently resulted schedule delay. In case of buried piles, it was never thought that some kind of parallel trench along with intense rain can create problem of uneven side support distribution which further resulted in the failure. Had it been envisaged, designer may restrict the affecting construction boundary along the pipe which would save cost and time. However, such interactions and failures are never conventional and these are the lessons learned in hard way. While designing the side slopes for the reservoir and containment structures, designer primarily focus on the side slopes considering slope stability as primary verification. However, steep side slopes, which may be stable for slope stability, may not be suitable for other construction elements like liners followed by protection tiles. In most of the cases veneers stability assessments are ignored which results mostly in the maintenance cost throughout the life cycle of the reservoir and containment structures.

Following key lessons can be attributed from the observed failures:

- For deep excavation and dewatering activities, correct assessment of earth pressures and toe stability are equally important. While grabbing from underwater, dewatering conditions and supporting arrangement shall be verified as it may trigger catastrophic situations.
- It is recommended to dewater from the outside of the excavation geometry if possible and careful monitoring at every stage will be helpful in avoiding any unforeseen failures.
- For buried pipes, imposed loading conditions require pipe analysis to be performed for different site conditions, operation conditions and scenario of operations. Parallel trench excavation along the pipeline is special conditions and execution team shall share all such relevant situations during the execution so that such damage to the already in place elements can be avoided.
- The construction sequence of buried pipes is long enough, so as not to ignore hydrostatic uplift due to floatation. It shall be verified in agreement with construction sequence, i.e., the absence of sufficient earth cushion and pipe running empty. If required, additional stabilizing elements may be included in the system design.
- The "veneer stability" of the liner protection elements is very important and must not be ignored. While following standard specifications, engineering performance shall be addressed first.
- Dyke resting on the pervious ground is unsafe condition and must not be ignored. If possible sufficient measures shall be taken during project inception as remedial measure can invite considerable cost and schedule overruns.
- The structures adjacent to the stockpile can be subjected to later pressure due to rapid and uncontrolled filling of stock pile especially of iron ore (having high unit weight). The subsoil receiving the stock shall be strong enough to prevent any lateral spreading and sliding and shall be improved if required to safeguard the surrounding structures, i.e., reclaimer foundations.
- Excavation footprint of deep buried structures and pipelines must be considered while planning the structures around the buried elements. The trench excavations and backfilling is very important construction sequence and shall be addressed very carefully. The tilting and settlement may create situations where constructed structures need to be abandoned thereby increasing national wastage of manpower and construction materials.

References

Shukla, J.C., K.M. Patel, N. Subbarao, K. Srinivas, and D.L. Shah. 2009. Construction of underground scale pit at Essar steel—Hazira-dewatering and consequences. In *Indian Geotechnical Conference*, paper 80.

Shukla, J.C., J. Chokshi, and A. Singhavi. 2016. Behavior of large diameter buried pipes—A case study of parallel trench excavation. *ASCE Geochicago* 2016: 462–474.

Reliability-Based Design of Pile Foundations

Sumanta Haldar

1 Introduction

The significance of variability in the geotechnical engineering field is well documented and efforts are being made to develop a rational design methodology over the traditional factor of safety approach (Becker 1996). Pile foundations are important components of critical infrastructure like bridges, transmission towers, buildings and offshore structures, and are often subjected to various uncertain loads due to wind, wave, traffic, or seismic activities. The properties of soil vary considerably within geologically distinct layers that may affect the mechanical behavior of pile foundation to a significant extent (e.g., Griffiths and Fenton 2001). The design becomes more challenging when the inherent variability and random spatial variability of soil properties is considered.

Conventional design methods of pile foundations are based on traditional global factor of safety approach (e.g. IS 2911-1-2 2010). Major drawback of factor of safety based approach is that it mostly based on experience and judgment and cannot be unique (Phoon 1995). The reliability-based design is advantageous over the factor of safety approach because it is rational and consistent as the variability can be included and it is associated with performance indicators (Becker 1996; Kulhawy and Phoon 2002). Several international design standards (e.g., FEMA 1994; NRC 1995; CEN 2001; FHWA 2001; AASHTO 2007) have introduced the limit state design or load and resistance factor design (LRFD) as the accepted method of design for structures and foundations (Ellingwood and Tekie 1999). Efforts are also being made to introduce reliability or probability based design for pile foundations (FHWA 2001; NRC 1995).

S. Haldar (✉)
Department of Civil Engineering, School of Infrastructure,
Indian Institute of Technology, Bhubaneswar, India
e-mail: sumanta@iitbbs.ac.in

K. Ilamparuthi and R. G. Robinson (eds.), *Geotechnical Design and Practice*,
Developments in Geotechnical Engineering,
https://doi.org/10.1007/978-981-13-0505-4_20

Pile foundations under static and dynamic loads are designed using two limit state functions, namely ultimate limit state and serviceability limit states (Phoon et al. 2003). Uncertainty due to fatigue is associated with the pile foundations of offshore wind turbine structure, hence fatigue limit state is also an essential design criterion (Carswell et al. 2015). Performance of a structure is evaluated using probability of failure or dimensionless parameter reliability index considering all limit states (Baecher and Christian 2003). Computation of probability of failure or reliability index is dependent on joint probability density of load and resistance. Evaluation of joint probability density function and evaluation of multiple integral for large numbers of random variables is practically difficult. Therefore, several analytical methods such as first-order reliability methods (FORM), second-order reliability methods (SORM) and simulation based methods such as Monte Carlo Simulation (MCS) approach are adopted to compute failure probability (Ang and Tang 1984). Currently, geotechnical community considers Load Resistance Factor Design (LRFD) approach to overcome the probabilistic computation demands statistical knowledge of the designer. In routine pile design LRFD approach is advantageous because development of reliability-based design approach does not require probabilistic computation, which are often applied to the pile foundation to account for the variability in load and resistance (Phoon and Kulhawy 2005). The FHWA (2001) manual recommends the use of LRFD for pile foundations supporting bridge structures as it provides a consistent level of safety in design.

Development of reliability-based design methodology of pile foundation is relevant and important in the context of current design paradigm. This paper presents an overview of reliability-based pile design methodology subjected to static and dynamic loads. The key elements of reliability-based design methods are emphasized and probabilistic design concepts of vertically and laterally loaded pile and offshore foundation are outlined.

2 Assessment of Reliability

Harr (1996) defined reliability as the probability of a system performing its required function adequately for a specified period of time under stated condition. The probability of failure is defined as the probability of an unwanted uncertain event and its unwanted consequences (CIRIA 2001).

The probability of failure of a structure is quantified using a specific performance criteria. In a system, if load is X and resistance is Y, the objective of reliability analysis is to ensure the event $X < Y$ for a specified life period. The performance function is defined as,

$$Z = g(X, Y) = Y - X \quad (1)$$

The limit state is defined as, $Z = 0$ which is the boundary between safe and unsafe regions as illustrated in Fig. 1.

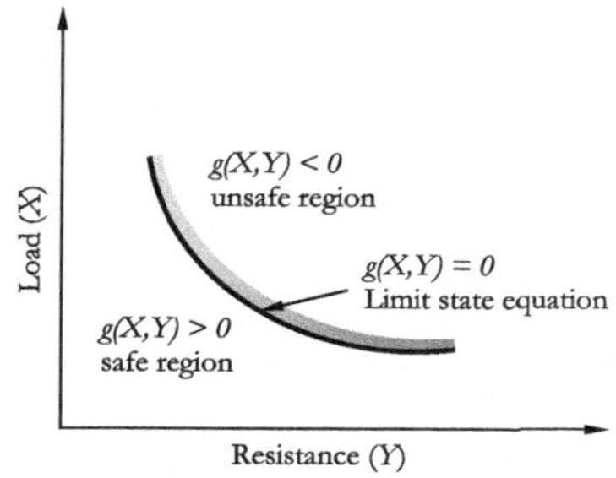

Fig. 1 Limit state concept (Haldar and Mahadevan 2000)

If X and Y are the independent random variables, $f_X(X)$ and $f_Y(Y)$ are probability density functions respectively, and $F_X(X)$, $F_Y(Y)$ are known cumulative distribution functions, the probability of failure is (Ang and Tang 1984),

$$p_{\mathrm{f}} = \int_0^{\infty} F_Y(x) f_X(x) \cdot \mathrm{d}x = \int_0^{\infty} [1 - F_X(y)] f_Y(y) \cdot \mathrm{d}y \tag{2}$$

The probability of failure can also be determined from joint probability density function of X and Y, $f_{XY}(X, Y)$ as (Haldar and Mahadevan 2000),

$$p_{\mathrm{f}} = \iint_{g(X,Y)<0} f_{XY}(X, Y) \cdot \mathrm{d}x \cdot \mathrm{d}y \tag{3}$$

Estimation of failure probability is dependent on $f_{XY}(X, Y)$. However, evaluation of joint probability density function and multiple integral for large numbers of random variables is difficult. Hence, the following methods are used to compute probability of failure or reliability index.

2.1 First-Order Reliability Method (FORM)

The first-order reliability method is based on first-order Taylor series expansion approximation. The performance function is linearized at mean values of the random variables. Considering X and Y are statistically independent random variables which follows normal distribution, the Cornell reliability index (β_{cornell}) is defined as (Cornell 1969),

$$\beta_{\mathrm{cornell}} = \frac{\mu_Z}{\sigma_Z} = \frac{\mu_Y - \mu_X}{\sqrt{\sigma_Y^2 + \sigma_X^2}} \tag{4}$$

where μ_Z and σ_Z are the mean and standard deviation of the performance function respectively, μ_X and σ_X are the mean and standard deviation of load respectively, and μ_Y and σ_Y are the mean and standard deviation of resistance respectively. The failure probability can be determined from reliability index as $p_f = 1 - \Phi(\beta_{\text{cornell}})$, where $\Phi(.)$ is the standard normal cumulative distribution function.

Cornell's reliability indices are not the same for equivalent performance functions. Hasofer and Lind (1974) represented reliability index as the shortest distance from the origin to the limit state surface when the surface is represented in the standard normal space. Hasofer-Lind reliability index (β_{HL}) can be expressed as,

$$\beta_{\text{HL}} = \min_{G(x)=0} \sqrt{(x')^t(x')} \tag{5}$$

where X_i' is the random variable with zero mean and unit standard deviation and the reduced variable is defined as $X_i' = \left(X_i - \mu_{X_i}\right)/\sigma_{X_i}$. The reduced limit state equation can be represented as $g(X') = 0$.

2.2 Monte Carlo Simulation (MCS)

In case of complex nonlinear problems, explicit performance functions are not available. Hence, real structures are analyzed using numerical analysis, e.g., finite element analysis and finite difference analysis. In such cases, FORM cannot be used to find reliability index or probability of failure in the limit state. The Monte Carlo simulation based approaches overcome these problems. The MC simulation is a repetitive simulation process where a particular set of values of random variables are generated with the known probability distributions and the outcome is presented in the form of histograms. It may be noted that each simulation process is deterministic. This method is computationally expensive, however considering the advancement in recent computer technology, Monte Carlo method is versatile and mature enough to be used for reliability analysis of large engineering structures (Marek and Haldar 2003).

The probability of failure for a limit state equation is quantified based on certain failure criteria. For a limit state function $g(.)$, failure occurs when $g(.) < 0$. If N is the total number of Monte Carlo simulations and N_f is the number of simulations such that $g(.) < 0$, the probability of failure is expressed as, $p_f = N_f/N$.

2.3 Load Resistance Factor Design (LRFD)

In LRFD, the resistance and loads are related as

$$(\text{RF})R_n \geq \sum_i (\text{LF})_i L_{n(i)} \tag{6}$$

where R_n is the nominal (design) resistance or capacity, $L_{n(i)}$ is the ith nominal load, RF is the resistance factor, $(LF)_i$ is the load factor corresponding to the ith load. The nominal loads are calculated by the design engineer deterministically from the weight of the superstructure and load-related coefficients (e.g., wind or seismic coefficient) based on codes or manuals. For pile foundations, vertical and lateral loads arise due to superstructure, traffic movement, seismic activities, wind flow, and wave motion. Considering a combination of dead load (DL) and live load (LL), the limit state function can be written as,

$$g(R, \mathrm{DL}, \mathrm{LL}) = R - \mathrm{DL} - \mathrm{LL} \tag{7a}$$

where R is the resistance (capacity). The FORM can be used to obtain the most probable failure point (i.e. design point) on the limit state surface in the reduced-variable probability space of R, DL, and LL for a target reliability index β_T (or, for a target probability of failure p_{fT}). For a given soil profile and pile geometry, the mean value of R is constant; therefore, most probable failure point depends on the applied loads. The most probable failure point represents a combination of resistance, dead load and live load, R^*, DL*, and LL* respectively, that has the maximum probability of causing a failure in the system (the 'distance' from the 'mean' point, corresponding to the mean values of R, DL and LL to the limit state surface in the reduced-variable probability space is minimum at the most probable failure point). In order to move away from the most probable failure point to the safe zone, resistance, and load factors are used:

$$R^* - \mathrm{DL}^* - \mathrm{LL}^* = 0 = (\mathrm{RF})R_n - (\mathrm{LF})_{\mathrm{DL}}(\mathrm{DL})_n - (\mathrm{LF})_{\mathrm{LL}}(\mathrm{LL})_n \tag{7b}$$

$$\mathrm{RF} = R^*/R_n = \lambda_M R^*/\mu_R \tag{7c}$$

$$(\mathrm{LF})_{\mathrm{DL}} = \mathrm{DL}^*/(\mathrm{DL})_n = \lambda_{\mathrm{DL}}\mathrm{DL}^*/\mu_{\mathrm{DL}} \tag{7d}$$

$$(\mathrm{LF})_{\mathrm{LL}} = \mathrm{LL}^*/(\mathrm{LL})_n = \lambda_{\mathrm{LL}}\mathrm{LL}^*/\mu_{\mathrm{LL}} \tag{7e}$$

where μ_R, μ_{DL} and μ_{LL} are the mean resistance, dead load, and live load respectively; λ_M, λ_{DL} and λ_{LL} are the bias factors for resistance, dead load, and live load respectively.

2.4 *Target Reliability Index or Probability of Failure*

The objective of reliability-based design is to ensure probability of failure shall not exceed an acceptable value (p_{fT}) or reliability index must not fall below the acceptable limit (β_T). Typical ranges of β_T and p_{fT} for pile foundation is summarized in Table 1.

Table 1 β_T or p_{fT} for pile foundations

Foundation	β_T	p_{fT}	Remarks
Drilled shaft for transmission line structures (Phoon et al. 1995)	3–3.2	1.3×10^{-3}–6.9×10^{-4}	Uplift load
	3–3.2	-do-	Compression
	2.9–3	1.3×10^{-3}	Lateral
General foundations (Becker 1996)	3.5	2.3×10^{-4}	Normal failure consequence
	4	3.2×10^{-5}	Severe failure consequence
Driven pile	2–2.5	2.3×10^{-2}–6.2×10^{-3}	NCHRP (2004)
Drilled shaft	2.5–3.5	6×10^{-3}–2.3×10^{-4}	
Axially loaded pile	2.3–2.8	10^{-2}–6×10^{-3}	Phoon et al. (1995)
Pile group consists of 5 piles or less	2.33	10^{-2}	Paikowsky (2002)
Pile group consists of more than 5 piles	3	1.4×10^{-3}	
Monopile for offshore wind turbine	3.7	1.1×10^{-4}	DNV-OS-J101 (2010)

3 Uncertainties in Pile Design

3.1 Inherent Uncertainty

Due to geological processes, soil properties vary in vertical and horizontal directions which is defined as inherent uncertainty of soil. Inherent uncertainty can be divided into two components, namely trend and fluctuation component. The inherent uncertainty of soil properties is modeled as a random field, which is described by the mean, coefficient of variation (COV) (= standard deviation/mean) and scale of fluctuation of soil properties. The scale of fluctuation is the measure of the distance of separation at which two samples are considered reasonably correlated (Vanmarcke 1983). Small values of scale of fluctuation denote rapid fluctuation of soil properties about the mean and large values indicate smoothly varying property. Ranges of inherent uncertainty of soil parameters (COV) e.g. undrained shear strength (s_u), angle of internal friction (ϕ), bulk unit weight (γ), SPT-N and relative density (D_r) are reported in Phoon and Kulhawy (1999). For example, COV of s_u varies from 20 to 55% for clay. Range of COV of ϕ, γ SPT-N and D_r are 5–15, 0–10, 25–50 and 50–70% respectively. Range of vertical and horizontal scale of fluctuation for s_u are 0.8–6.1 m and 50 m respectively.

3.2 *Model Uncertainty*

Model uncertainty arises because of the mathematical models and equations used in engineering analysis which are not perfect representations of physical processes but approximations of the actual behavior based on a set of assumptions. Model uncertainty in numerical model (e.g., *p–y* based approach) for predicting pile responses can be quantified from field load test results. The pile responses can be obtained from numerical model and matched with the corresponding responses from the large number of field tests. The predicted and measured pile responses can be compared. The predictive capability of the numerical model is evaluated by defining a model bias factor which is the ratio of measured and predicted responses. The mean of bias factor represents the tendency of the numerical model to overestimate or underestimate the responses and the COV of bias factor reflects the model uncertainty.

Estimation of model uncertainty from field tests has limitations because there are several sources of uncertainty associated with field measurements (e.g., inherent soil variability and measurement error) and it is very difficult to isolate the model uncertainty (Phoon and Kulhawy 2005). However, the measurement uncertainty and inherent spatial variability of soil are minimized by high quality tests and averaging effect of soil properties, respectively.

3.3 *Measurement Uncertainty*

Measurement uncertainty arises during the estimation of soil properties from the equipment, procedural-operator and random testing effects. Phoon and Kulhawy (1999) indicated that the equipment effect normally arises from the inaccuracies in the measuring devices, procedural-operator incurred limitations in the existing standard test and the way they are followed. The range of measurement uncertainty for SPT, CPT, triaxial and direct shear test are 15–45, 5–15, 7–56 and 3–29% respectively (Phoon and Kulhawy 1999).

4 Reliability-Based Design of Pile Foundation

Implementation of reliability-based design of pile foundation is presented with reference to few examples.

4.1 *Effect of Spatial Variability of Soil*

A probabilistic study on single pile foundation subjected to vertical and lateral loads has been conducted by Haldar and Babu (2008, 2012). Effect of soil spatial

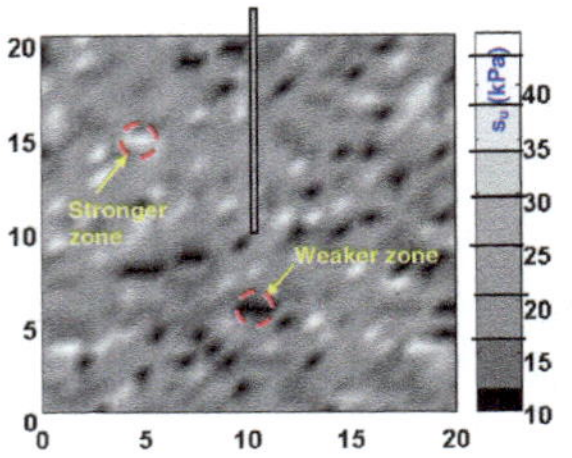

Fig. 2 Typical realization of random field, COV of s_u (COV_{su}) = 30% and scale of fluctuation (δ_z) = 5 m (Haldar and Babu 2008)

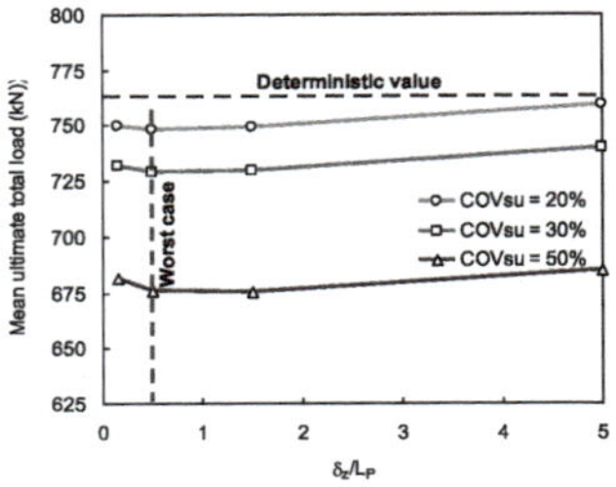

Fig. 3 Effect of spatial variability on mean total ultimate vertical capacity (Haldar and Babu 2012)

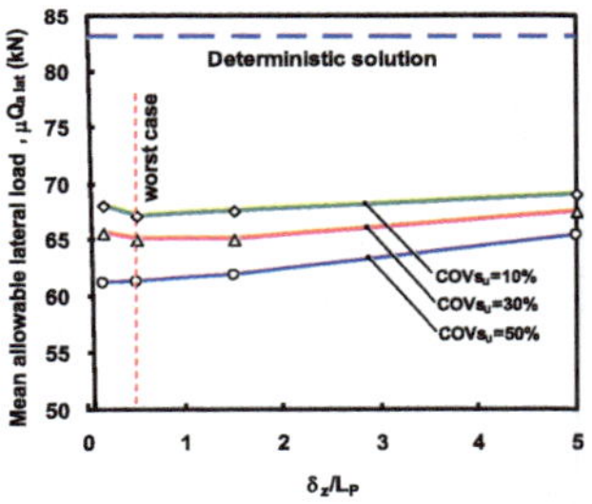

Fig. 4 Effect of spatial variability on mean allowable lateral capacity (Haldar and Babu 2008)

variability on ultimate vertical capacity and allowable lateral capacity is evaluated. A 2D log-normal homogeneous random field is generated (Fig. 2) in numerical model using *Cholesky* decomposition method. Undrained shear strength of soil (s_u) is considered as random variable and assumed to be a log-normally distributed random field represented by mean, standard deviation and scale of fluctuation of s_u. It is further assumed that the soil has an isotropic Markovian correlation structure. This makes the scale of fluctuation the same in both the horizontal and vertical directions. Monte Carlo simulation is used to determine the probabilistic characteristics of the ultimate vertical and allowable lateral capacity.

Figure 3 shows that substantial reduction of mean ultimate vertical capacity was observed due to the effect of soil spatial variability that that of case of deterministic results. The reduction in pile capacity was observed to be maximum at $\delta_z/L_p = 0.5$ (L_p = pile length), which is considered to be worst scale of fluctuation. Similar trend was also observed for laterally loaded pile as shown in Fig. 4. It was observed that, for spatially varied soil, the resulting mean allowable lateral load corresponding to allowable lateral deflection of 0.05 m at pile head is significantly lower than that of the case of deterministic soil. It is interesting to note that the weak zones in each realization of random field are useful to understand the variations in the failure mechanism as well as vertical or lateral capacity of pile. This aspect is important for pile design.

4.2 LRFD of Laterally Loaded Pile

Design of laterally loaded piles are mostly guided by the serviceability limit state of excessive lateral deflection. Haldar and Basu (2014) proposed resistance factors for laterally loaded piles embedded in clay deposits in which the soil properties are assumed to be random variables. Uncertainties associated with the response of laterally loaded piles was taken into account based on the serviceability limit state of excessive lateral deflection. Lateral load carrying capacity of pile were determined based on a specified allowable lateral deflection at the pile head. The pile load-displacement curves were generated using the *p–y* analysis. Model uncertainty in *p–y* model has been incorporated in the analysis. The uncertainties associated with the lateral capacity, for a specified lateral head deflection, were quantified. Probability distribution of the lateral capacity was determined using Monte Carlo simulations.

Uncertainty in *p–y* model (API 2011) parameters (ε_{50}: the strain corresponding to one half of the maximum stress in laboratory undrained compression tests performed on undisturbed clay samples and *J*: dimensionless empirical coefficient) and model uncertainty are quantified based on 40 field pile load test data. Two parameters were back calculated from the numerically generated load-deflection curves with the field curves by trial and error. The allowable capacities corresponding to allowable deflection (δ_a) = 15 and 25 mm were obtained from the fitted curve and field load-deflection curve. The statistics of bias factors (ratio of measured and predicted capacities) were obtained. Uncertainty in ε_{50} and *J* are presented in Fig. 5.

The resistance factors were obtained using the first-order reliability method (FORM) for a given target p_f, uncertainties in s_u (COV_{su} = 20%, log-normal distribution), dead load (COV_{DL} = 10%, normal distribution), live load (COV_{LL} = 25%, log-normal distribution), lateral capacity (COV = 20%) and model uncertainties (COV = 15%). The resistance factors (RF) for different DL/LL ratios and target p_f are presented in Fig. 6.

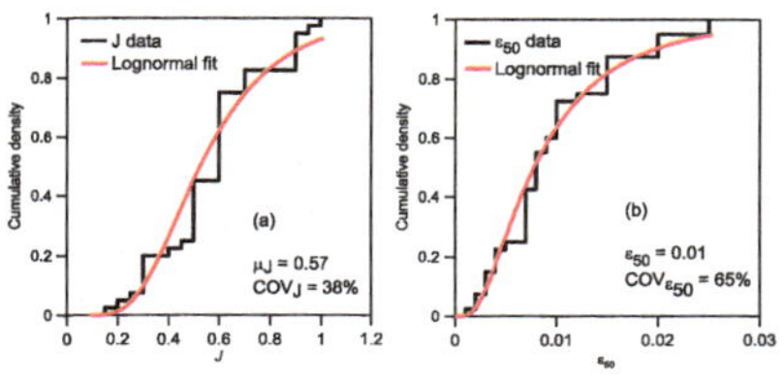

Fig. 5 Uncertainty of *J* and ε_{50} from field load test results (Haldar and Basu 2014)

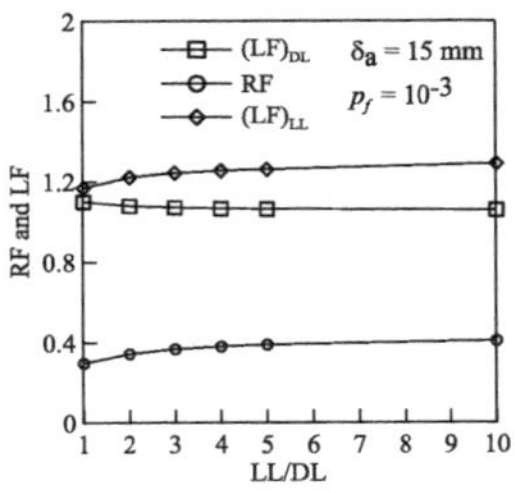

Fig. 6 RF and LF for different DL/LL ratios

4.3 Reliability of Monopile Foundation for Offshore Wind Turbine

Haldar and Basu (2016) assessed the reliability monopile-supported offshore wind turbine (OWT) considering the uncertainties in aerodynamic and hydrodynamic loads and spatial variability of soil. The study also includes climate change effect on future wind and wave loads. Hence, time-dependent reliability assessment has been performed. Because of excessive foundation cost (about 30–40% of the total cost), a rational and robust foundation design strategy is required. The dynamic response of the OWT was estimated using finite element analysis. Soil resistance was modeled using the API (2011) based *p–y* curve. Spatial variability of angle of internal friction ($COV_{\phi'} = 5\%$, $\delta_v = 2$ m) is mapped onto *p–y* curves. Statistical downscaling model was used to forecast future wind speed, wave height and wave period. Aerodynamic and hydrodynamic loads were estimated based on DNV-OS-J101 (2010). The properties of a 5 MW OWT was selected from the National Renewable Energy Laboratory (NREL 2004). A time-dependent reliability assessment was carried out considering limit state of serviceability (SLS).

Results are presented in Fig. 7 for different values of time span. Considering climate change effects, the estimated time-dependent reliability index was found to be less than that of the case when climate change effect is not taken into account in the analysis.

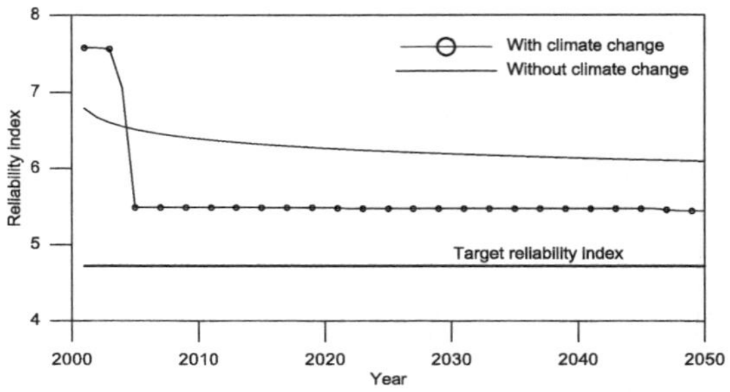

Fig. 7 Reliability of OWT considering SLS for present and future climate conditions for different time span (Haldar and Basu 2016)

5 Concluding Remarks

This paper shows a few example problems to elucidate significance of reliability-based design of pile foundation. Geotechnical engineers have realized the significance of variability in foundation design and an important step is taken to calibrate the design codes of practice considering variability using probabilistic model. There is a necessity to incorporate reliability analysis as theoretical basis for foundation design and development of simplified LRFD for routine design. It can be concluded that the reliability-based design of pile foundations considering variability and spatial correlation of soil enables a rational choice of design.

References

AASHTO. 2007 *AASHTO LRFD Bridge design specifications*, Washington, DC.

Ang, A.H.-S., and W.H. Tang. 1984. *Probability concepts in engineering planning and design. Volume II: Decision, risk and reliability*. New York: Wiley.

API. American Petroleum Institute. 2011. *Petroleum and natural gas industries-specific requirements for offshore structures. Part 4—Geotechnical and foundation design considerations*.

Baecher, G.B., and J.T. Christian. 2003. *Reliability and Statistics in geotechnical engineering*. London and New York: Wiley.

Becker, D.E. 1996. 'Eighteenth Canadian geotechnical colloquium: Limit states design for foundations, Part I. An overview of the foundation design process. *Canadian Geotechnical Journal* 33: 956–983.

Carswell, W., S.R. Arwade, D.J. DeGroot, and M.A. Lackner. 2015. Soil-structure reliability of offshore wind turbine monopile foundations. *Wind Energy* 18 (3): 483–498.

CIRIA. 2001. *RiskCom: Software tool for managing and communicating risks*. Construction Industry Research and Information Association, London.

Cornell, C.A. 1969. A probability-based structural code. *ACI Journal* 66 (12): 974–985.

DNV-OS-J101. 2010. *Design of offshore wind turbine structures*. DET NORSKE VERITAS.

Ellingwood, B.R., and P.B. Tekie 1999. Wind load statistics for probability-based structural design. *Journal of Structural Engineering, ASCE* 125 (4): 453–463.

European Committee for Standardization (CEN). 2001. *Eurocode 7 Part 1: Geotechnical design: General rules*, December 2001, Brussels, Belgium.

Federal Emergency Management Agency (FEMA). 1994. *NEHRP Recommended provisions for the development of seismic regulations for new buildings*. FEMA Rep. 223, Washington, D.C.

Federal Highway Administration (FHWA). 2001. *Load and resistance factor design (LRFD) for highway bridge substructures*. FHWA HI-98-032, FHWA, Washington, DC.

Griffiths, D.V., and G.A. Fenton. 2001. Bearing capacity of spatially random soil: The undrained clay Prandtl problem revisited. *Geotechnique* 51 (4): 351–359.

Haldar, A., and S. Mahadevan. 2000. *Probability, reliability and statistical methods in engineering design*. New York: Wiley.

Haldar, S., and D. Basu. 2014. Resistance factors for laterally loaded piles in clay. In *Proceedings of geo-congress 2014*, 3333–3342.

Haldar, S., and D. Basu. 2016. Effect of climate change on the reliability of offshore wind turbine foundations. In *Proceedings of geo-Chicago 2016*, 407–417.

Haldar, S., and G.L.S. Babu. 2008. Effect of soil spatial variability on the response of laterally loaded pile in undrained clay. *Computers and Geotechnics* 35 (4): 537–547.

Haldar, S., and G.L.S. Babu. 2012. Response of vertically loaded pile in clay: A probabilistic study. *Geotechnical and Geological Engineering* 30 (1): 187–196.

Harr, M.E. 1996. *Reliability-based design in civil engineering*. Mineola, New York: Dover Publications Inc.

Hasofer, A.M., and N.C. Lind. 1974. Exact and invariant second moment code format. *Journal of Engineering Mechanics, ASCE* 100: 111–121.

IS 2911-1-2. 2010. *Design and construction of pile foundations, Part 1: Concrete piles, section 2: Bored cast in-situ concrete piles*, Bureau of Indian Standards.

Kulhawy, F.H., and K.-K. Phoon. 2002. Observations on geotechnical reliability-based design development in North America. In *Proceedings of foundation design codes and soil investigation in view of international harmonization and performance*, IWS Kamakura, Tokyo, Japan.

Marek, P., and A. Haldar. 2003. Simulation as an alternative to codified risk-based engineering design. In *Proceedings of 4th international symposium on uncertainty modelling and analysis*, 222.

National Research Council of Canada (NRC). 1995. *National building code of Canada*. Ottawa: NRC.

NCHRP. 2004. *Load and resistance factor design (LRFD) for deep foundations*. Transportation Research Board, National Research Council, NCHRP report 507, Washington, D.C.

NREL. 2004. *National renewable energy laboratory*, Washington.

Paikowsky, S.G. 2002. *Load & resistance factor design (LRFD) for deep foundations*. NCHRP Report 24-17, Transportation Research Board, Washington.

Phoon, K.K., F.H. Kulhawy, and M.D. Grigoriu. 1995. *Reliability-based design of foundations for transmission line structures*. Report TR-105000, Electric Power Research Institute, Palo Alto.

Phoon, K.-K. (1995) *Reliability-based design of foundations for transmission line structures*. Ph. D. Thesis, Cornell University, Ithaca, New York.

Phoon, K.-K., and F.H. Kulhawy. 1999. Evaluation of geotechnical property variability. *Canadian Geotechnical Journal* 36: 625–639.

Phoon, K.-K., and F.H. Kulhawy. 2005. Characterisation of model uncertainties for laterally loaded rigid drilled shafts. *Geotechnique* 55 (1): 45–54.

Phoon, K.-K., F.H. Kulhawy, and M.D. Grigoriu. 2003. Development of a reliability-based design framework for transmission line structure foundations. *Journal of Geotechnical and Geoenvironmental Engineering, ASCE* 129 (9): 798–806.

Vanmarcke, E.H. 1983. *Random fields: Analysis and synthesis*. Cambridge: MIT Press.

Large Deformation Modelling in Geomechanics

Krishna Kumar and Kenichi Soga

1 Introduction

Geophysical hazards and industrial processes involves flow of dense granular material. Understanding the mechanics of granular flow is of particular importance in predicting the run-out distances of debris flows. The dynamics of a homogeneous granular flow involve at least three distinct scales: the microscopic scale, which is characterised by contact between grains, the meso-scale that represents micro-structural effects such as grain rearrangement, and the macroscopic scale, where geometric correlations can be observed (Cambou et al. 2009). Different approaches have been adopted to model granular flows at different scales of description. Conventionally, granular materials such as soils are modelled as a continuum. On a macroscopic scale, granular materials exhibit many collective phenomena and the use of continuum mechanics to describe the macroscopic behaviour can be justified. However, on a grain scale, the granular materials exhibit complex solid-like and/or fluid-like behaviour depending on how the grains interact with each other. Recent works on granular materials suggest that a continuum law may be incapable of revealing in-homogeneities at the grain-scale level, such as orientation of force chains, which are purely due to micro-structural effects (Topin et al. 2011). Discrete Element approaches are capable of simulating the granular material as a discontinuous system allowing one to probe into local variables such as position, velocities, contact forces, etc. The fundamental question is how to model granular materials which exhibit complex phenomenon.

K. Kumar (✉)
Department of Engineering, University of Cambridge, Cambridge, UK
e-mail: kks32@cam.ac.uk

K. Soga
Department of Civil and Environmental Engineering,
University of California, Berkeley, USA
e-mail: soga@berkeley.edu

K. Ilamparuthi and R. G. Robinson (eds.), *Geotechnical Design and Practice*,
Developments in Geotechnical Engineering,
https://doi.org/10.1007/978-981-13-0505-4_21

The collapse of a granular column on a horizontal surface is a simple case of granular flow, however a proper model that describes the flow dynamics is still lacking. Experimental investigations have shown that the flow duration, the spreading velocity, the final extent of the deposit, and the energy dissipation can be scaled in a quantitative way independent of substrate properties, grain size, density, and shape of the granular material and released mass (Lajeunesse et al. 2005; Lube et al. 2005). Simple mathematical models based on conservation of horizontal momentum capture the scaling laws of the final deposit, but fail to describe the initial transition regime. Granular flow is modelled as a frictional dissipation process in continuum mechanics but the lack of influence of inter-particle friction on the energy dissipation and spreading dynamics (Lube et al. 2005) is surprising.

In order to describe the mechanism of saturated and/or immersed granular flows, it is necessary to consider both the dynamics of the solid phase and the role of the ambient fluid (Denlinger and Iverson 2001). In two-phase models (Pitman and Le 2005), the momentum transfer between the grains and the suspension fluid depends on the momentum equations of both phases. Using mixture theory-based models, the shear-induced migration and grains collisions can be considered in an average sense. However, the dynamics of the solid phase alone are insufficient to describe the mechanism of submerged granular flow in fluid. In particular, when the solid phase reaches a high volume fraction, the strong heterogeneity arising from the contact forces between the grains and the hydrodynamic forces, is difficult to integrate into the homogenization process involving global averages such as the mixture theory.

In the present study, multi-scale numerical modelling, i.e. discrete element and continuum analyses, of the quasi-two-dimensional collapse of granular columns are performed using Discrete Element (DEM) approach and Material Point Method (MPM) to understand the ability and limitations of continuum approaches in modelling dense granular flows. The effect of fluid on the run-out behaviour of a granular column collapse is studied using a 2D lattice Boltzmann-Discrete Element method. The flow kinematics are compared with the dry and buoyant granular collapse to understand the influence of hydrodynamic forces and lubrication on the run-out behaviour.

2 Numerical Set-Up and Configuration

The granular column collapse experiment involves filling a column of height H_0 and length L_0 with a granular material of mass 'm'. The granular column is then released *en masse* by quickly removing the gate, thus allowing the granular material to collapse onto the horizontal surface, forming a deposit having a final height H_f and length L_f. A computational study on an equivalent two-dimensional configuration (circular discs) was carried out using Discrete Element and Continuum (MPM) approaches.

3 Discrete Element Approach

Two-dimensional DEM analyses were carried out by varying the aspect ratio ('a') of the granular column from 0.2 to 10. The contact models are well established in a 2D discrete element approach, which enables us to understand the mechanism of energy dissipation in dense granular flows. The normal contact force is modelled using linear-spring with dashpot. The tangential contact force is modelled using a sliding/striking tangential friction model (Luding 2008). An inter-particle friction coefficient of 0.53 is adopted. Further details about the micro-mechanical parameters used in this study are presented in Soundararajan (2015).

4 Material Point Method

Material Point Method (MPM) (Bardenhagen et al. 2000; Sulsky et al. 1994) is a particle based method that represents a body as a collection of material points, and their deformations are determined by Newton's laws of motion. MPM is a hybrid Eulerian–Lagrangian approach, which uses moving material points, and computational nodes on a background mesh. This approach is very effective particularly in the context of large deformations (Soga et al. 2016). Although, not derived directly from what is classically considered as mesh-free or mesh-less methods, the background mesh is used only to perform discretization, integration, and to solve equations of motions.

The Material Point Method involves discretizing a domain with a set of material points and assigning an initial value of position, velocity, mass, volume and stress. The material points are assumed to be within the computational grid, which is assumed to be a Cartesian lattice for convenience. At every time step t_k, the MPM computation cycle involves projecting the data, such as position, mass and velocity from the material points to the computational grid using standard nodal basis functions, called shape functions, derived based on the position of particle with respect to the grid. Gradient terms are calculated in the computational grid and the governing equation, i.e. the equation of motion, is solved and the updated position and velocity values are mapped back to the material points. The mesh is reinitialized to its original state and the computational cycle is repeated. The MPM solution scheme is shown in Fig. 1.

In the MPM simulations, the granular flow was assumed to be in the critical state and the critical state friction angle was used as an input in the Mohr–Coulomb model. A bi-periodic shear test using DEM grains yielded a critical state friction angle of 23°.

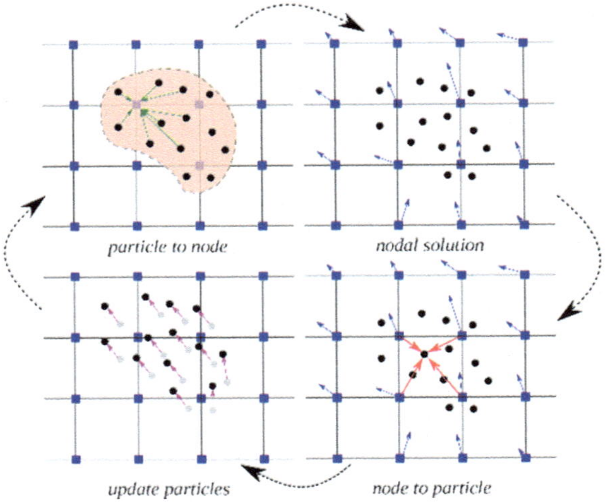

Fig. 1 MPM solution scheme

5 Flow Kinematics and Energy Dissipation

The final run-out distance normalised to the initial length is shown in Fig. 2 as a function of the initial aspect ratio. Two distinct flow regimes can be observed: (a) for '*a*' < 1.7 a linear relation between the spread and aspect ratio can be observed, and (b) for '*a*' > 1.7 a power-law relationship exists. MPM and DEM simulations are able to capture the linear relationship for '*a*' < 1.7, and the simulation results agree with the experimental investigation (Lajeunesse et al. 2005). Significant difference between MPM, which is based on a simple frictional model for dissipation of potential energy, and DEM simulations for '*a*' > 1.7 indicates a change in the mechanism of energy dissipation for columns with large aspect ratios ('*a*' > 1.7). A constant frictional dissipation model cannot describe a power-law

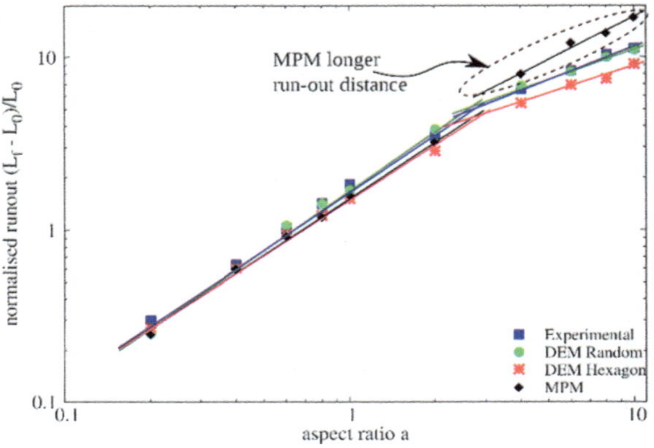

Fig. 2 Evolution of run-out with initial aspect ratio

relation (Staron and Hinch 2007). A transition in the run-out behaviour at an aspect ratio of 1.7 indicates a change in flow dynamics.

The flow dynamics of two granular columns (‘a’ = 0.4 and ‘a’ = 6) are compared to understand the difference in the energy dissipation mechanism, and in turn on the run-out behaviour for different aspect ratios. The granular column collapse is initiated by a failure at the edge of the pile along a well-defined failure surface. The grains located above the failure surface move “en masse” leaving a static region underneath the failure surface. After a transient time of order τ_c, defined as $(H_o/g)^{0.5}$, the flow is fully developed and the flow comes to a rest at time $t = 3\ \tau_c$ (Staron and Hinch 2007). Columns with small aspect ratio fail by avalanching of the flanks, forming a truncated cone-like deposit. Velocity profiles of a granular column with ‘a’ = 0.4 are presented in Fig. 3. Both MPM and DEM simulations show a well-defined failure surface and predict similar run-out behaviour. Although, the MPM simulation predicts 25% longer for the flow to be fully mobilised (see Fig. 3), the flow evolution happens quicker, which can be attributed to the marginally lower mobilised potential energy. The initial potential energy stored in the column is converted to kinetic energy that is then dissipated as the material flows down. MPM and DEM predict similar energy evolution behaviour which indicates that most of the stored initial potential energy is dissipated through friction. Hence, for columns with smaller aspect ratios, the run-out distance is proportional to the mass flowing above the failure.

A significant difference in the run-out behaviour between MPM and DEM (see Fig. 2) indicates that a simple frictional model is insufficient to describe the flow kinematics for columns with large aspect ratios. Velocity profiles of a granular column with ‘a’ = 6 are presented in Fig. 4. At critical time ($t = \tau_c$), both MPM and DEM predict almost identical run-out behaviour on a distinct failure plane. The flow evolution is much faster in MPM and results in a larger run-out distance. For larger aspect ratios, the flow is still initiated by a well-defined failure surface. However, the centre of gravity of the granular column is much higher than the top of the failure surface, which results in free fall of grains under gravity consuming the column along their way. When they reach the vicinity of the failure surface, the grains undergo collisions with the bottom plane and the neighbouring grains, thus causing the flow to deviate along the horizontal direction releasing a large amount of kinetic energy gained during the free fall (see Fig. 5). The grains then eventually leave the base area of the column and flow sideways undergoing frictional dissipation. The process involves collective dynamics of all the particles, and DEM simulations model both collisional and frictional dissipation process. However, MPM simulations assume that the total initial potential energy stored in the system is completely dissipated through friction over the entire run-out distance, resulting in larger run-out distance.

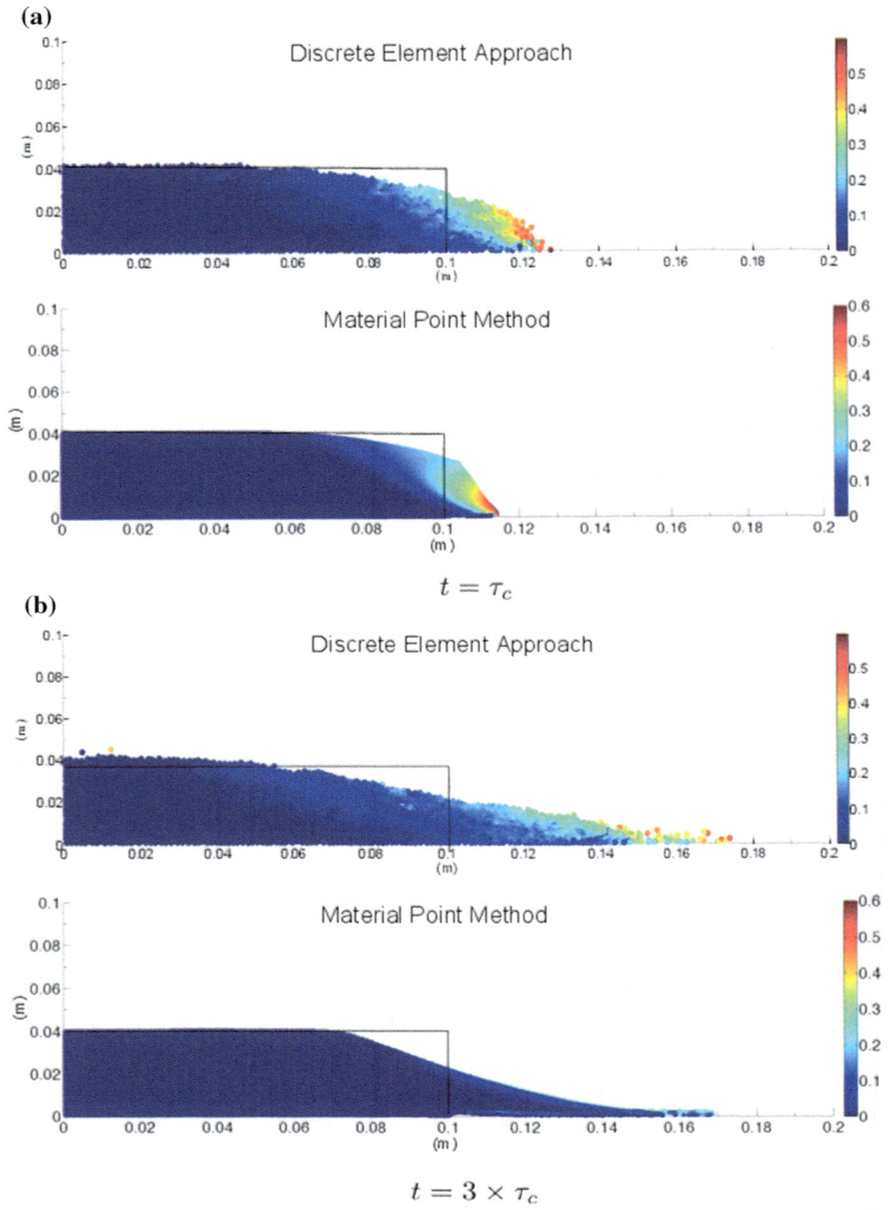

Fig. 3 Comparison of flow profile between DEM and MPM of a column with 'a' = 0.4

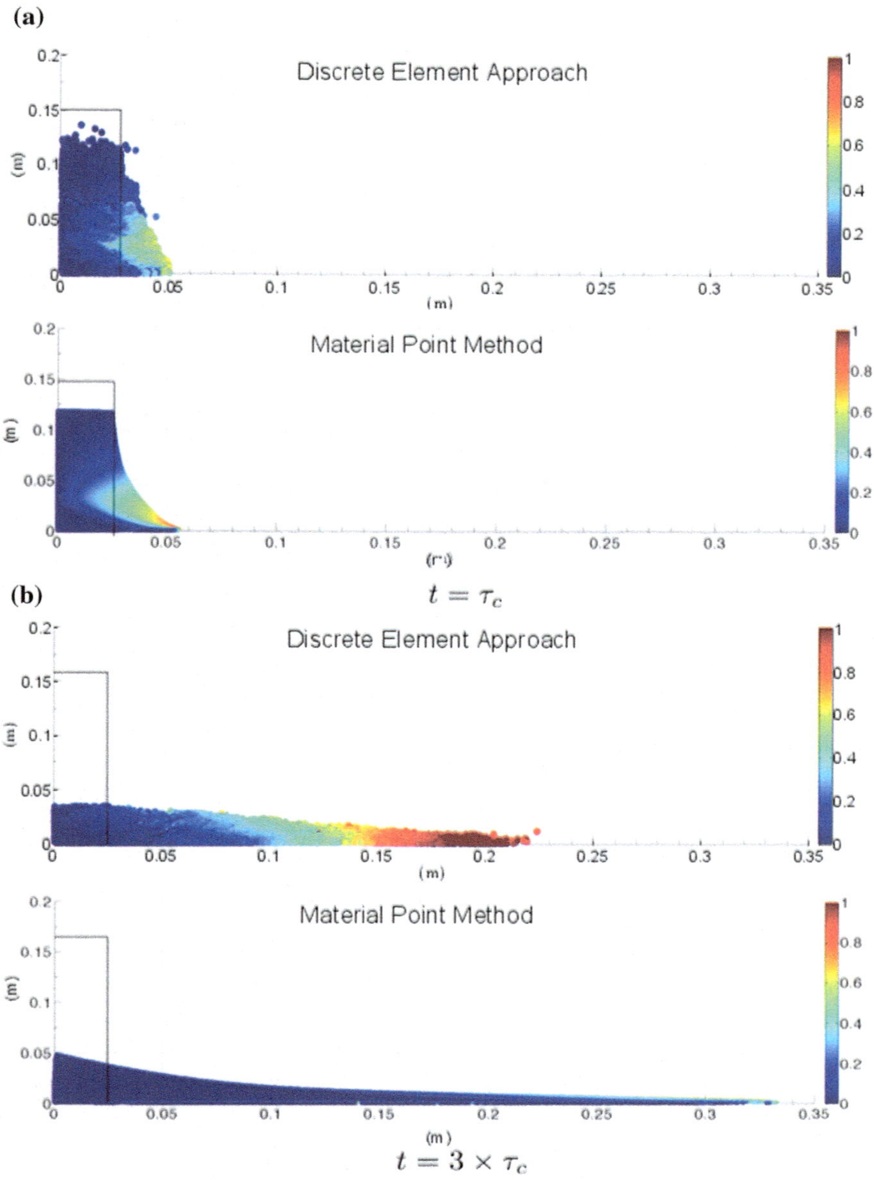

Fig. 4 Comparison of flow profile between DEM and MPM of a column with '*a*' = 6

Fig. 5 Energy evolution for a column with '*a*' = 6

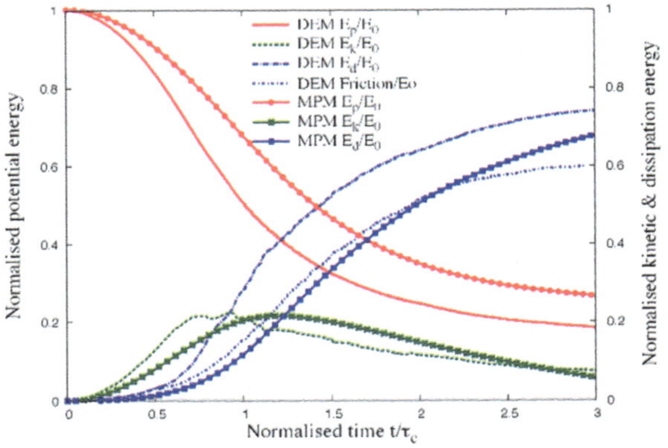

6 LBM–DEM Formulation

The Lattice Boltzmann equation Method (LBM) is an alternative approach to the classical Navier–Stokes solvers for fluid flow and works on an equidistant grid of cells, called lattice cells, which interact only with their direct neighbours (He and Luo 1997). The fluid domain is divided into a rectangular grid or lattice, with the same spacing 'h' in both the x- and the y-directions, as shown in Fig. 6. The present study focuses on 2-D problems; hence, the D2Q9 momentum discretization is adopted, where the fluid particles at each node are allowed to move to their eight intermediate neighbours with eight different velocities $\mathbf{e}_i$ (i = 1, …, 8).

The nine discrete velocity vectors are defined as

$$\begin{cases} e_0 = (0,0) \\ e_1 = C(1,0); \quad e_2 = C(0,1); \quad e_3 = C(-1,0); \quad e_4 = C(0,-1); \\ e_5 = C(1,1); \quad e_6 = C(-1,1); \quad e_7 = C(-1,-1); \quad e_8 = C(1,-1); \end{cases} \tag{1}$$

in which C is the lattice speed (= h/Δt). Where, Δt is the discrete time step. The primary variables in the Lattice Boltzmann formulation are called the fluid density distribution functions, f_i, each relating the portable amount of fluid particles moving with the velocity e_i along the ith direction at each node. The macroscopic variables such as density and velocity can be obtained from the particle distribution

Fig. 6 The Lattice Boltzmann discretization and D2Q9 scheme: **a** a standard LB lattice; **b** D2Q9 model

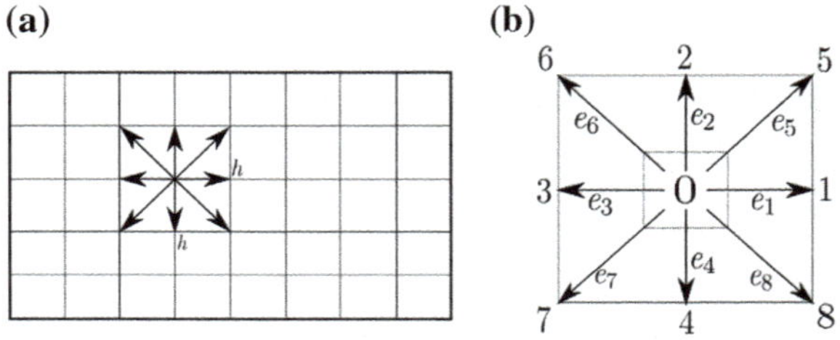

functions. There are nine fluid density distribution functions, f_i ($i = 0, \ldots, 8$), associated with each node in the D2Q9 model. This simple equation allows us to pass from the discrete microscopic velocities that comprise the LBM back to a continuum of macroscopic velocities representing the fluid's motion. The BGK (Bhatnagar-Gross-Krook) approximation is used to describe the streaming and the collision of the particles. Streaming and collision (i.e. relaxation to local equilibrium) is defined as

$$f_i(\mathbf{x}+\mathbf{e}_i\Delta t, t+\Delta t) = f_i(\mathbf{x}, t) - \frac{1}{\tau}[f_i(\mathbf{x}, t) - f_i^{eq}(\mathbf{x}, t)] \quad (i = 0, \ldots, 8),$$

where the left-hand side describes the streaming part and the collision part, which brings the system to local equilibrium is described by the right hand side. τ is a non-dimensional relaxation time parameter, which is related to the fluid viscosity.

6.1 LBM–DEM Coupling

Lattice Boltzmann approach can accommodate large grain sizes and the interaction between the fluid and the moving grains can be modelled through relatively simple fluid–grain interface treatments. Further, employing the Discrete Element Method (DEM) to account for the grain–grain interaction naturally leads to a combined LB–DEM procedure (Kumar et al. 2012). The Eulerian nature of the LBM formulation, together with the common explicit time step scheme of both LBM and DEM makes this coupling strategy an efficient numerical procedure for the simulation of grain–fluid systems. Such a coupled methodology is used in simulating grain–fluid systems dominated by grain–fluid and grain–grain interactions. To capture the actual physical behaviour of the fluid–grain system, it is essential to model the boundary condition between the fluid and the grain as a non-slip boundary condition, i.e. the fluid near the grain should have similar velocity as the grain boundary. The solid grains inside the fluid are represented by lattice nodes. The discrete nature of lattice, results in a stepwise representation of the surfaces, which are circular, hence sufficiently small lattice spacing is adopted.

In a 3D granular assembly, the pore spaces between the grains are interconnected, whereas in a 2-D assembly, a non-interconnected pore-fluid space is formed as the grains are in contact with each other. This means that the fluid enclosed between the grains cannot flow to the neighbouring pore spaces; this results in an unnatural no flow condition in a 2-D case. In order to overcome this effect, a reduction in radius is assumed only during the LBM computation (fluid and fluid–solid interaction) steps. The hydrodynamic radius r, allows for interconnected pore space through which the pore-fluid can flow thus simulating a 3D-flow-like behaviour. The reduction in the radius is assumed only during LBM computations, hence this technique has no effect on the grain–grain interactions computed using DEM.

7 Granular Avalanches in Fluid

In this section, we study the behaviour of immersed granular avalanches for different permeability. We consider 2D poly-disperse system ($d_{max}/d_{min} = 1.8$) of circular discs in fluid. The simulations were carried out with 1000 grains of density 2650 kg/m^3 and a contact friction angle of 26°. The collapse of the column was simulated inside a fluid with a density of 1000 kg/m^3 and a kinematic viscosity of 1×10^{-6} m^2/s. The choice of a 2D geometry has the advantage of cheaper computational effort than a 3D case, making it feasible to simulate very large systems with an important number of nodes for a reasonable computing time. A granular column of aspect ratio 'a' of 0.8 was considered. Radius of the grains was varied from 0.7 to 0.95R during LBM computations. Dry and buoyant analyses were done to compare the effect of hydrodynamic forces on the run-out distance.

The final run-out profile for a reduced grain size of 0.7 and 0.95R are presented in Fig. 7. The less permeable column runs out further, while the highly permeable column entraps more water and has smaller run-out distance. The rheology of the flow tends to change, with change in permeability. It can be observed from the figure that the dry and buoyant columns run farther and evolve quicker than the immersed case. The run-out is found to increase with decrease in permeability. The time required for the flow to initialize increases with decrease in permeability, i.e. a highly permeable column collapse quicker in comparison to a less permeable granular column. Three distinct regimes can be observed in the immersed avalanche, similar to the observations in dry granular column collapse: (1) a vertical-fall regime, where the grains located at the top undergo vertical falls and the grains at the bottom are ejected horizontally by the fluid; (2) a heap regime, where the grains at the top move along an inclined stationary deposit; (3) a horizontal regime, where the grains move essentially horizontally. Figure 8 shows a high value of positive pore-pressure at the face of the column in the low permeable condition. This indicates that the low permeable column fails as a continuous block undergoing a shear failure, which generates a very large negative excess

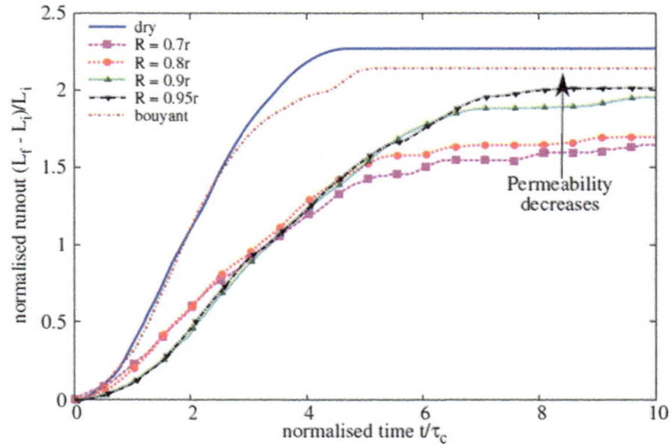

Fig. 7 Evolution of normalised run-out distance with time for different initial permeability

Fig. 8 Effect of permeability on the excess pore water pressure distribution at $t = 3\ \tau_c$ along the horizontal direction at a height of 10 d from the base

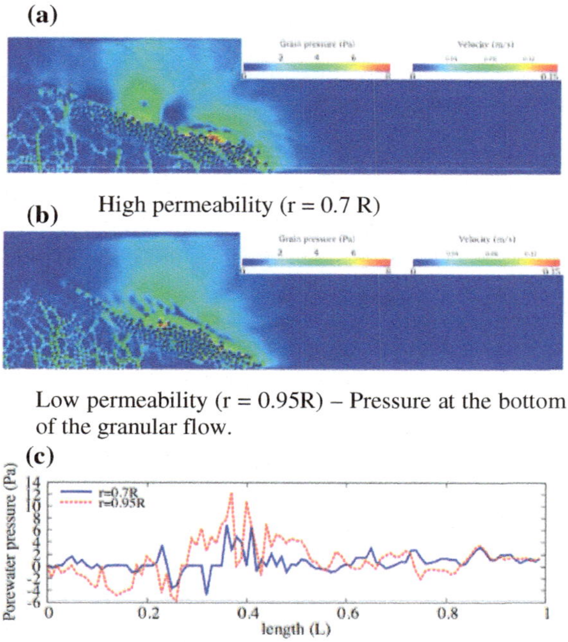

(a) High permeability (r = 0.7 R)

(b) Low permeability (r = 0.95R) – Pressure at the bottom of the granular flow.

(c) Porewater pressure at bottom plane for low and high permeability cases

pore-pressure along the failure surface. As the material flows, it entraps water and creates a lubrication layer resulting in farther run-out distances. However, in the case of the high permeable column, the failure is more localised with multiple negative excess pore-pressure spikes.

8 Summary

Multi-scale simulation of granular column collapse was performed to understand the ability and limitations of continuum models to capture the micro-mechanics of dense granular flows. The run-out behaviour predicted by both continuum and DEM simulations matches for columns with small aspect ratios, where the dissipation is predominantly frictional. However, MPM predicts larger run-out distances for columns with higher aspect ratios. Energy evolution studies using DEM simulations reveal that the run-out behaviour is independent of frictional properties of the granular material and collision predominates the initial free-fall regime. The lack of a collisional energy dissipation mechanism in MPM results in over prediction of run-out distances. LB-DEM simulations were performed to understand the behaviour of submarine granular flows. The run-out distance in fluid increases with decrease in permeability. The run-out distance increases with decreasing permeability. An increase in the hydrodynamic radius from 0.7 to 0.95R increases the normalised run-out by 25%. For the same value of peak kinetic energy, the run-out distance in fluid is longer than the dry column collapse. Also, with decreasing permeability the run-out distance increases for the same peak kinetic energy. The low permeability of the granular mass results in entrainment of water causing hydroplaning. With decreasing permeability, the duration for the flow to initiate increases as the time required to dissipate the large negative excess pore-pressures increases.

References

Bardenhagen, S.G., J.U. Brackbill, and D. Sulsky. 2000. The material-point method for granular materials. *Computer Methods in Applied Mechanics and Engineering* 187 (3–4): 529–541.

Cambou, B., M. Jean, and F. Radjaï. 2009. *Micromechanics of granular materials*. Wiley-ISTE.

Denlinger, R.P., and R.M. Iverson. 2001. Flow of variably fluidized granular masses across three-dimensional terrain, ii: Numerical predictions and experimental tests. *Journal Geophysical Research* 106 (B1): 553–566.

He, X., and L.-S. Luo. 1997. Theory of the lattice Boltzmann method: From the Boltzmann equation to the lattice Boltzmann equation. *Physical Review E* 56 (6): 6811–6817. https://doi.org/10.1103/PhysRevE.56.6811.

Kumar, K., K. Soga, and J.-Y. Delenne. 2012. Granular flows in fluid. In *Discrete element modelling of particulate media*, ed. C.-Y. Wu. Cambridge: Royal Society of Chemistry. http://doi.org/10.1039/9781849735032.

Lajeunesse, E., J.B. Monnier, and G.M. Homsy. 2005. Granular slumping on a horizontal surface. *Physics of Fluids* 17 (10).

Lube, G., H.E. Huppert, R.S.J. Sparks, and A. Freundt. 2005. Collapses of two-dimensional granular columns. *Physical Review E—Statistical, Nonlinear, and Soft Matter Physics* 72 (4): 1–10.

Luding, S. 2008. Cohesive, frictional powders: Contact models for tension. *Granular Matter* 10 (4): 235–246. https://doi.org/10.1007/s10035-008-0099-x.

Pitman, E.B., and L. Le. 2005. A two-fluid model for avalanche and debris flows. *Philosophical Transactions: Mathematical, Physical and Engineering Sciences* 363 (1832): 1573–1601.

Soga, K., E. Alonso, A. Yerro, K. Kumar, and S. Bandara. 2016. Trends in large-deformation analysis of landslide mass movements with particular emphasis on the material point method. *G{é}otechnique* 66 (3): 248–273. article.

Soundararajan, K.K. 2015. *Multi-scale multiphase modelling of granular flows*. Ph.D. thesis, University of Cambridge.

Staron, L., and E.J. Hinch. 2007. The spreading of a granular mass: Role of grain properties and initial conditions. *Granular Matter* 9 (3–4): 205–217.

Sulsky, D., Z. Chen, and H.L. Schreyer. 1994. A particle method for history-dependent materials. *Computer Methods in Applied Mechanics and Engineering* 118 (1–2): 179–196.

Topin, V., F. Dubois, Y. Monerie, F. Perales, and A. Wachs. 2011. Micro-rheology of dense particulate flows: Application to immersed avalanches. *Journal of Non-Newtonian Fluid Mechanics* 166 (1–2): 63–72.

Geotechnical Engineering Through Mind Maps

M. Rama Rao and P. Samatha Chowdary

1 Introduction

The Mind Map is an expression of radiant thinking and imagination and is therefore a natural function of the human mind. It is a powerful graphic technique which provides a universal key for unlocking the potential of the brain. In its simplest definition, a mind map is a diagram that visually represents information said. In another way, it maps the information in your mind generally a particular subject. The real power of Mind Map is persons own ability to convey information quickly and easily. The term Mind Map was first popularized by British psychologist Tony Buzan in 1960s.

1.1 Literature Review

Buzan and Buzan (1993) stated that during learning process, the human brain primarily remembers the following:

- Items from the beginning of the learning period ("the primacy effect")
- Items from the end of the learning period ("the recency effect")
- Any items associated with the things or patterns already stored, or linked to other aspects of what is being learned.

M. Rama Rao (✉) · P. Samatha Chowdary
R.V.R & J.C.CE, Guntur 522019, AP, India
e-mail: ramarao.muvvala@gmail.com

P. Samatha Chowdary
e-mail: samathaponduri@gmail.com

K. Ilamparuthi and R. G. Robinson (eds.), *Geotechnical Design and Practice*,
Developments in Geotechnical Engineering,
https://doi.org/10.1007/978-981-13-0505-4_22

- Any items which are emphasized as being in some way outstanding or unique.
- Any items which appeal particularly strongly to any of the five senses.
- Those items which are particular interest to the person.

This information paved the way for development of mind mapping which enhances vibrant thinking abilities incorporating a field of different colors.

Buzan (2002) in his book on "How to Mind Map" stated that a mind map is the easiest way to put information into our brain and to take information out of our brain. And also mentioned that it is a creative and effective means of note-taking and also maps out our vibrant thoughts.

Mento et al. (1999) declared that Mind Mapping is a powerful cognitive tool which can be used in a variety of ways because of its ability to evoke associative and nonlinear thinking.

Sperry (1968) stated that Mind Mapping takes into account that the two halves of the brain (right and left) where in which the left brain deals with naming and categorizing things, language, reading, writing, arithmetic, and symbolic deduction. While the right brain operates in a non-verbal way (using images) and excels in dealing with emotions, forms and patterns, colors, rhythm, perceptual and intuitive information. The right brain processes information quickly in a nonlinear and non-sequential style.

Al-Jarf (2009) investigated the impact of using Mind Mapping software on "Enhancing Freshman students" acquisition of English writing skills.

The findings revealed that the written work produced by students using Mind Mapping included more relevant detail and better organized with connected ideas than the work of the control group. It was also found that Mind Mapping raised the performance of students at all levels of ability as they became more efficient in generating and organizing ideas for their writing. The students also displayed a positive attitude towards using Mind Mapping as a pre-writing activity.

Boyson (2009) stated that, the visibility of Mind Mapping provides an effective approach for promoting better understanding in students as a pedagogical tool. Its flexibility also means that it possesses several uses in the classroom. Using Mind Mapping for lesson planning can help teachers identify a logical teaching route and increases recall of the subject matter. This can boost teaching confidence and facilitate the smooth running of lessons.

Researchers Goodnough and Woods (2002) in their valuable founding's stated that Mind Mapping provides to be an effective and useful strategy for introducing new concepts, providing a whole-class focus for a large research project, assess student learning and offer greater choice to students to complete their assignments and projects.

Mento et al. (1999) observed that a number of executive students made clear and compelling presentations using only a transparency of their Mind Map, without fumbling about with notes. These students were also able to handle challenging questions with confidence. Their ability to handle the presented material in such an effective way was attributed to better recall of the information because it had been captured and stored in an integrated, radiating manner rather than linearly.

Mind Mapping was selected as an effective strategy for encouraging creativity development in engineering students and was applied with encouraging results. Zampetakis et al. (2007) stated that Mind map allowed creativity development to be introduced into the curriculum in a way that made best use of the time available.

1.2 The Disadvantages of Standard Notes Making in Teaching

Important ideas are conveyed by key words; generally they are nouns or strong verbs. In standard notes these key words are often appear on different pages, obscured by the mass of less important words. These factors prevent the brain from making appropriate associations between key words.

In linear notes we use words, lines, numbers, logic, and sequence. They represent brain's "left-brain" skills and do not include any of right-brain skills, which enable us to comprehend rhythm, color, and space, and to daydream. Students around the world are taught to take notes in blue/black inks. Monotonous (single color) notes are visually boring. Standard notes often take the form of endless similar looking lists making it almost impossible to remember their content.

By encouraging unnecessary noting, reading, and re-reading of unnecessary notes, searching for key words waste time in all stages not only that linear presentation of standard notes prevents the brain from making associations, thus counteracting creativity and memory.

1.3 Applications of Mind Map in Teaching

Mind maps can be used in numerous practical ways to make teaching and learning easier and more enjoyable. It can be used for the following purposes.

1.3.1 Preparing Lecture Notes

One of the most powerful ways to use mind maps is lecture notes. Preparing a lecture in mind map form is faster than writing it out and has the advantage of allowing teacher and student to keep on overview of the subject. Mind map also enables the speaker to have a perfect balance between a spontaneously spoken and fresh talk, on one hand, and the clear and well-structured presentation on the other. It allows accurate time—keeping during the lecture.

1.3.2 Lesson Preparation

Teaching preparation can amount to files full of plans and documents, as a single lesson's preparation alone can generate reams of notes. By using Mind Mapping for planning, the amount of notes can be reduced and the total plans will be converted to clear—concise plans that are easy to follow.

1.3.3 Delivering Lessons

Mind Maps are ideal for teaching and presenting concepts in the classroom as they provide a useful focus for students by delivering an overview of the topic without superfluous information. Mind maps have been proven to be perfect for introducing a new subject in a way which is accessible and easy to follow.

1.3.4 Curriculum Planning

Teaching today hinges around well-organized curriculum planning. With new curriculum initiatives, briefs, objectives and aims, it can all become overwhelming. But with Mind Mapping sequence can be well planned and key topics can be prioritized by adding deadlines as necessary.

1.3.5 Creating Handouts

Mind Maps are the ideal teaching tool for classroom handouts, as the inherent color, images and visually appealing layout of a Mind Map engages students instantly. Mind Maps provide subject overviews succinctly, making even the most complex topic easy to understand and interesting.

1.3.6 Encouraging Discussion and Independent Thought

Mind Mapping is the perfect collaborative tool for class discussions, as the nature of the Mind Map encourages students to forge links between topics as well help in forming their own ideas and opinions.

1.3.7 Student Assessment

It is important to assess knowledge at the beginning of a topic and after to monitor the students understanding. Mind Mapping is a key tool for this concept, of preview and review (pre and post learning). Mind Maps encourages students to express ideas, from special needs and highly gifted students, and provides an accurate barometer of topic adoption.

1.3.8 Self-evaluation

With frequent teacher inspections and reviews, it is important to constantly self-assess and evaluate your teaching style and preparation. Mind Mapping helps in assessing current abilities of a person (for example, in areas such as lesson delivery, handouts, interaction, etc.) also help in setting goals, a person would like to achieve within a week, a month, or year. This powerful form of self-evaluation will allow teachers in improving and meeting targeted teaching objectives.

1.4 Mind Mapping Process

Mind maps are the reflection of your brain's natural image filled thinking processes and abilities. Mind map work—images with networks of association.

Mind mapping makes it easy to add information because we add additional ideas simply by inserting additional branches. Mind mapping written content requires systemized approach of **Preview–Read–Mind Map–Reviews** explained below.

Preview: Process in which we are prepared for what we are about to read leading valuable starting point for your mind map.
Read: Process in which the text is read from beginning to end and how the information can be organized into a mind map is set.
Mind Map: After previewing and reading the material, mind mapping is started with the title of the book, article or chapter in the center or with a key word that best sums up the idea.
Review: After mind map is created, it is compared to the material which have been read earlier in read.

Steps to Create Mind Map
A mind map simply requires these three, Blank unlined paper, colored pens, very important our imagination.

Mind map generally starts with a central idea or focus, which is the theme of the map. Related ideas branch out from the center. A sample mind map (Fig. 1) look like this.

A detailed insight of various steps to create a mind map is given below.

(1) Start in the center of a blank page turned sideways. Starting in the center gives brain freedom to spread out in all directions and to express itself more freely and naturally.

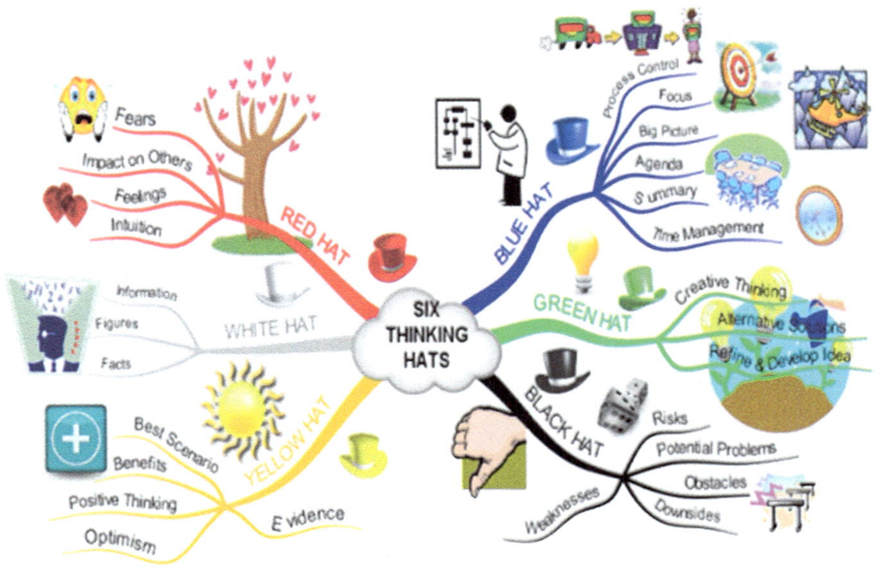

Fig. 1 Simple mind map

(2) Use an image or picture for your central idea.
(3) Use colors throughout. Colors are as exciting as to brain as are images. Color adds extra vibrancy and life to mind map, adds tremendous energy to creative thinking, and is fun.
(4) Connect main branches to the central image and connect second- and third-level branches to the first and second levels. Because brain works by association. Connecting the branches, makes understanding and remembrance easy.
(5) Make branches curved rather than straight-lined. As they are attractive and riveting your eye.
(6) Use one key word or image per line only. It generates its own special array of associations and connections.
(7) Use images throughout. Because each image, like central image, is also worth thousand words.

1.5 Advantages of Usage of Mind Map Software

The latest software can allow teachers and students to draw a mind map on screen. The advantages of this are obvious. Mind maps information saved in a file can be easily transferred to others. Paykoç et al. (2004) stated that use of Mind Mapping software improved student's freedom to reflect, make changes and develop relationships in line with their thinking process. Al-Jarf (2009) proved that Mind Mapping software offers a powerful approach for improving the ability of students to generate, visualize, and organize ideas. The students involved reported that the Mind Mapping tool encouraged creative thinking and they became faster at generating and organizing ideas for their writing.

Computer mind maps also allows us to store vast amounts of data in Mind Map form, to cross-reference that data to shift branches around from one part of mind map to another and rearrange entire mind maps in light of new information.

"iMindMap" is a software that gives a professional edge to a creative and fun process. It is a fantastic tool for teaching, having wide range of applications as attachment of files, videos, documents and web links to Maps for information at our fingertips providing an option to edit later. Helps in planning of essays that are especially useful for students for facing timed essays in their exams with notes in place.

"iMindMap" software also helps in organizing teaching plans even easier, as all the notes, files and education website links from within one Mind Map, also helps in planning presentation with all of our ideas and resources, also by allowing to present using the amazing inbuilt Presentation Mode.

iMindMap also assists in recording Audio Notes of ours and our fellow students, provokes discussions, adding each person's input to their individual branch.

iMindMap makes the process of curriculum Planning even more hassle-free through integration with Office software programs, such as Microsoft Word and Excel. And also the iMindMap Ultimate Project Management system gives us even more control over our annual plans and deadlines.

2 An Example on Geotechnical Engineering Topic

Relative density an aspect of geotechnical engineering is a term generally used to describe the degree of compaction of coarse-grained soils.

2.1 Preparation of Notes by Standard Note Making Method

By following standard note making method study of relative dry density is generally done as given below.

(1) **Relative Density or Density Index (D_r or I_D):**

It is most important soil aggregate property of coarse-grained soil. Relative density is a term generally used to describe the degree of compaction of coarse-grained soils. Relative density or density index is the ratio of the difference between the void ratios of a cohesion less soil in its loosest state and existing natural state to the difference between its void ratio in the loosest and densest states.

$$\text{Relative density} = D_r = \frac{e_{max} - e}{e_{max} - e_{min}} \times 100, \tag{1}$$

where

e_{max} void ratio of coarse-grained soil in its loosest state.
e_{min} void ratio of coarse-grained soil in its densest state.
e void ratio of coarse-grained soil in its natural existing state in the field.

It is generally expressed in percentage. When natural soil is in loosest in state, $D_r = 0\%$ and it is 100% in the densest state. Loose granular soil or low D_r Value is unstable under vibratory loads. Soil engineers are qualitatively describes the granular soil in terms of relative densities, as shown in Table 1.

It is also expressed in terms of dry unit weights are as follows:

$$e = \frac{G\gamma_w}{\gamma_d} - 1$$
$$e_{max} = \frac{G\gamma_w}{\gamma_{d(min)}} - 1$$
$$e_{min} = \frac{G\gamma_w}{\gamma_{d(max)}} - 1$$

Substituting e_{max}, e_{min} and e in Eq. (1),

$$\begin{aligned} D_r &= \frac{\left[\frac{1}{\gamma_{d(min)}}\right] - \left[\frac{1}{\gamma_d}\right]}{\left[\frac{1}{\gamma_{d(min)}}\right] - \left[\frac{1}{\gamma_{d(max)}}\right]} \times 100 \\ &= \left[\frac{\gamma_d - \gamma_{d(min)}}{\gamma_{d(max)} - \gamma_{d(min)}}\right]\left[\frac{\gamma_{d(max)}}{\gamma_d}\right] \times 100 \end{aligned} \tag{2}$$

Since the volume of soil of a given dry mass is proportional to (1 + e), density index can also be expressed as

Table 1 Relative density of granular soils

Relative density (%)	Classification
<15	Very loose
15–35	Loose
35–65	Medium
65–85	Dense
>85	Very dense

$$D_r = \frac{V_{max} - V}{V_{max} - V_{min}}, \tag{3}$$

where

V volume in natural state
V_{max} maximum volume of soil mass
V_{min} minimum volume of soil mass.

Porosity of soil depends on shape of grain, uniformity of grain size and condition of sedimentation. Hence porosity of soil does not indicate whether the soil is loose or dense. Relative density is better indicator when compared to porosity or void ratio. Two sand deposits possessing same grain shape and size but engineering behavior is different because of difference in relative density.

Determination of relative density is helpful in compaction of coarse-grained soils and various soil vibration problems associated with earth works, pile driving, and foundations of machinery.

2.2 Preparation by Using iMindMap Software

From Sect. 2.1 in standard note making method it can be seen that definition is in one line, equation somewhere in the middle of notes, again some notes. When these notes is read again for preparation creates monotony and at times confusion of what have been studied earlier there by making remembrance difficult. And even presentation of the lecture in this way makes it difficult to memorize.

Hand drawn mind map, after reading different conditions and to imagination we can see only key words are written in mind map. It also projects all the equations in single page to the equations in standard methodology of learning in different pages. This enables students to handle difficult equations with more ease. Since Mind Map possess more vibrant colors that are easy to memorize.

From Sect. 2.2 usage of iMindMap software makes the essence of the study simpler by making into little fragments and then connecting. In Fig. 2 Mind Map is shown with only key words by using collapse function. This enables student to easily remember topics involved in it.

In Fig. 3 when we expand definition branch, only definition will be visible with relevant formulae. This enables the student easily understand the definition and equations written in one page makes remembrance of equations at a stage.

In Fig. 4 when we expand classification branch, tabular form relevant to classification will appear to make it ready to read the table and memorize.

Likewise different other branches can be expanded and seen individually. We get final mind map with all the information as seen in Fig. 5 in Sect. 2.2.

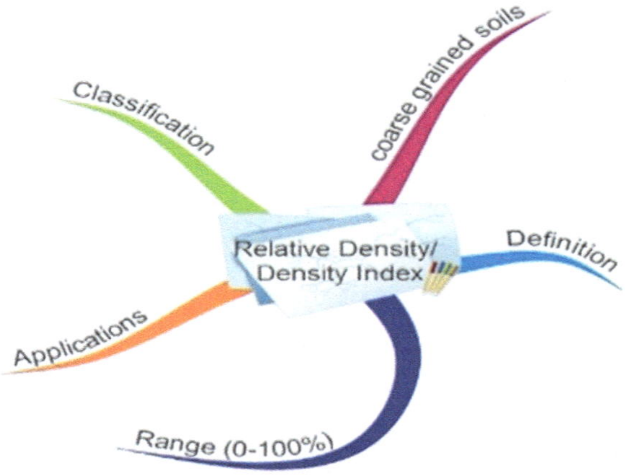

Fig. 2 Mind map showing key words

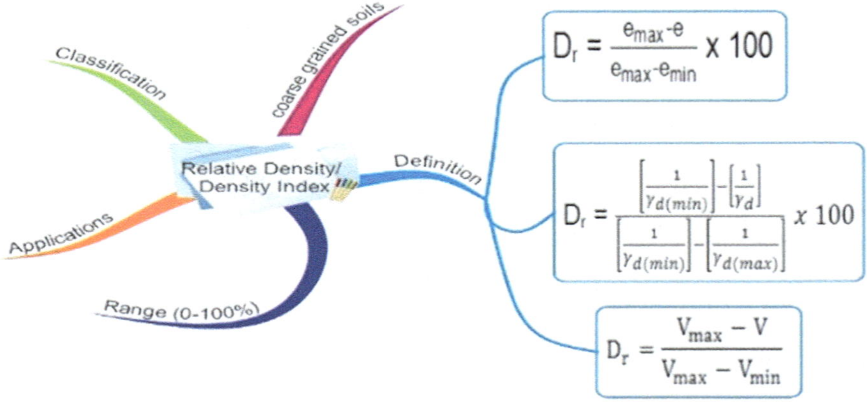

Fig. 3 Definition and formula

Thus iMindMap software can be used for dividing the total study into fragments by using collapse function to give an overall idea of study. Else the total picture and also individual key words can be studied in detail by using expand function. Not only that iMindMap software enables to share Mind Maps in variety of ways with others, by exporting to PDF, image files, a webpage and even MS Word and power point presentation. Mind Maps can be directly presented by pressing on the presentation mode available in iMindMap software.

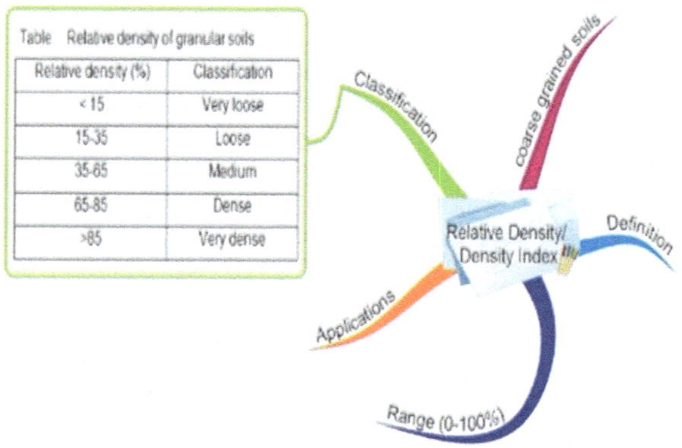

Fig. 4 Definition and classification

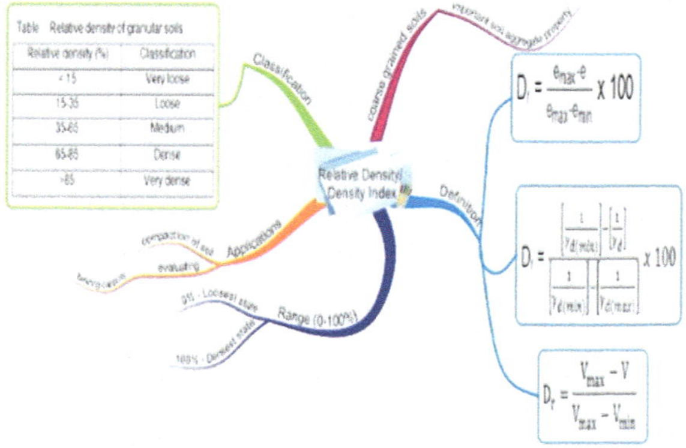

Fig. 5 Final mind map with all informations

3 Sample Mind Maps on Geotechnical Engineering Without Linear Notes

See Figs. 6 and 7.

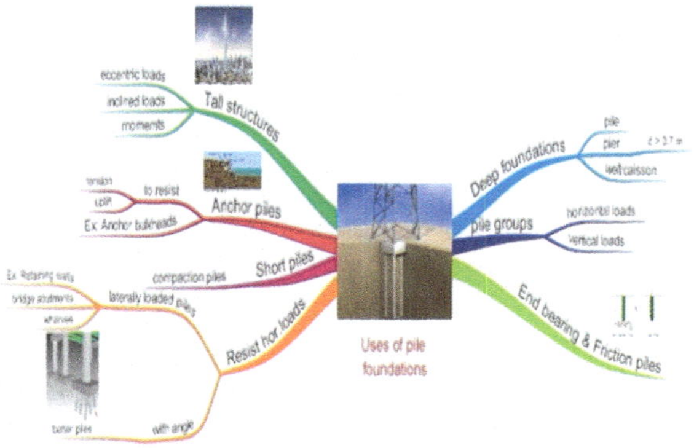

Fig. 6 Uses of pile foundation

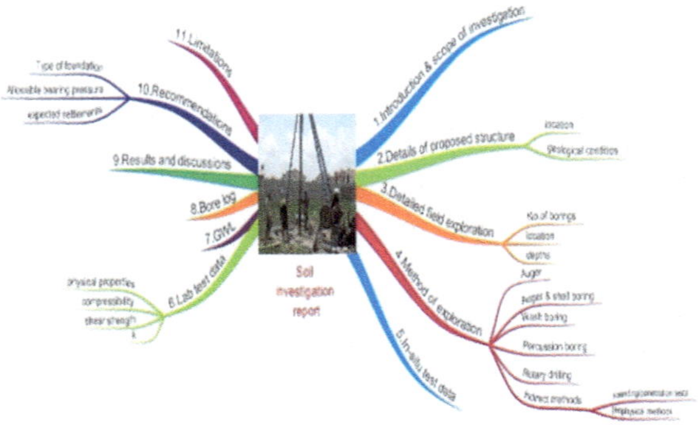

Fig. 7 Soil investigation report

4 Conclusions

- Inspire interest in the students.
- Mind maps converts teacher's notes to more flexible and adaptable.
- The physical volume of lecture notes is dramatically reduced.
- Mind maps present only relevant material in a clear and memorable form.
- Make Presentations more spontaneous, creative and enjoyable, both for teacher and students.
- Mind maps proves to be one of the best memorizing practices and also easy methodology to learn geotechnical engineering of many assumptions, equations and theory.

References

Al-Jarf, R. 2009. Enhancing freshman students' writing skills with a mind mapping software. Paper presented at the 5th International scientific conference, 375–382. eLearning and Software for Education, Bucharest, April 2009.

Boyson, G. 2009. *The use of mind mapping in teaching and learning*, 3. The Learning Institute, Assignment.

Buzan, Tony. 2002. *How to mind map.* Harper Collins publishers limited.

Buzan, Tony, and Barry Buzan. 1993. *The mind map book: How to use radiant thinking to maximize your brain's untapped potential.* USA: Penguin.

Goodnough, K., and R. Woods. 2002. Student and teacher perceptions of mind mapping: A middle school case study. Paper presented at the annual meeting of American Educational Research Association, 1–14, New Orleans, 1st to 5th April 2002.

Mento, A.J., P. Martinelli, and R.M. Jones. 1999. Mind mapping in executive education: Applications and outcomes. *The Journal of Management Development* 18 (4): 390–416.

Paykoç, F., B. Mengi, P.O. Kamay, P. Onkol, B. Ozgur, O. Pilli, and H. Yildirim. 2004. What are the major curriculum issues? The use of mind mapping as a brainstorming exercise. Paper presented at the first international conference on concept mapping, 293–296, Spain, 2004.

Sperry, R.W. 1968. Hemispheric deconnection and unity in conscious awareness. *Scientific American* 23: 723–735.

Zampetakis, L.A., L. Tsironis, and V. Moustakis. 2007. Creativity development in engineering education: The case of mind mapping. *Journal of Management Development* 26 (4): 370–380.